Reactions of Polyelectrolytes With Other Water Treatment Chemicals

Awwa Research Foundation

About the Awwa Research Foundation

The Awwa Research Foundation (AwwaRF) is a member-supported, international, nonprofit organization that sponsors research to enable water utilities, public health agencies, and other professionals to provide safe and affordable drinking water to consumers.

The Foundation's mission is to advance the science of water to improve the quality of life. To achieve this mission, the Foundation sponsors studies on all aspects of drinking water, including supply and resources, treatment, monitoring and analysis, distribution, management, and health effects. Funding for research is provided primarily by subscription payments from approximately 1,000 utilities, consulting firms, and manufacturers in North America and abroad. Additional funding comes from collaborative partnerships with other national and international organizations, allowing for resources to be leveraged, expertise to be shared, and broad-based knowledge to be developed and disseminated. Government funding serves as a third source of research dollars.

From its headquarters in Denver, Colorado, the Foundation's staff directs and supports the efforts of more than 800 volunteers who serve on the board of trustees and various committees. These volunteers represent many facets of the water industry, and contribute their expertise to select and monitor research studies that benefit the entire drinking water community.

The results of research are disseminated through a number of channels, including reports, the Web site, conferences, and periodicals.

For subscribers, the Foundation serves as a cooperative program in which water suppliers unite to pool their resources. By applying Foundation research findings, these water suppliers can save substantial costs and stay on the leading edge of drinking water science and technology. Since its inception, AwwaRF has supplied the water community with more than $300 million in applied research.

More information about the Foundation and how to become a subscriber is available on the Web at **www.awwarf.org**.

Reactions of Polyelectrolytes With Other Water Treatment Chemicals

Prepared by:
Audrey D. Levine
Department of Civil and Environmental Engineering
University of South Florida, Tampa, FL 33620-5350

Brian A. Bolto
CSIRO Manufacturing & Infrastructure Technology
PO Box 56, Highett, Vic 3190, Australia

and

David R. Dixon
Department of Chemical and Biomolecular Engineering
University of Melbourne, Parkville, 3010, Melbourne, Australia

Sponsored by:
Awwa Research Foundation
6666 West Quincy Avenue, Denver, CO 80235-3098

Published by:

Awwa Research Foundation
Advancing the Science of Water®

IWA Publishing

DISCLAIMER

Published by IWA Publishing, Alliance House, 12 Caxton Street, London SW1H 0QS, UK

Telephone: +44 (0) 20 7654 5500; Fax: +44 (0) 20 7654 5555; Email: publications@iwap.co.uk;
Web: **www.iwapublishing.com**

AwwaRF report number 91034F. Originally published by AwwaRF for its subscribers in 2004.
IWA Publishing version published 2005.

ISBN 1 84339 923 7

CONTENTS

TABLES

FIGURES

FOREWORD

The Awwa Research Foundation is a nonprofit corporation that is dedicated to the implementation of a research effort to help utilities respond to regulatory requirements and traditional high-priority concerns of the industry. The research agenda is developed through a process of consultation with subscribers and drinking water professionals. Under the umbrella of a Strategic Research Plan, the Research Advisory Council prioritizes the suggested projects based upon current and future needs, applicability, and past work; the recommendations are forwarded to the Board of Trustees for final selection. The Foundation also sponsors research projects through the unsolicited proposal process; the Collaborative Research, Research Applications, and Tailored Collaboration programs; and various joint research efforts with organizations such as the U.S. Environmental Protection Agency, the U.S. Bureau of Reclamation, and the Association of California Water Agencies.

This publication is a result of one of these sponsored studies, and it is hoped that its findings will be applied in communities throughout the world. The following report serves not only as a means of communicating the results of the water industry's centralized research program but also as a tool to enlist the further support of the nonmember utilities and individuals.

Projects are managed closely from their inception to the final report by the Foundation's staff and a large cadre of volunteers who willingly contribute their time and expertise. The Foundation serves a planning and management function and awards contracts to other institutions such as water utilities, universities, and engineering firms. The funding for this research effort comes primarily from the Subscription Program, through which water utilities subscribe to the research program and make an annual payment proportional to the volume of water they deliver and consultants and manufacturers subscribe based on their annual billings. The program offers a cost-effective and fair method for funding research in the public interest.

A broad spectrum of water supply issues is addressed by the Foundation's research agenda: resources, treatment, and operations, distribution and storage, water quality and analysis, toxicology, economics, and management. The ultimate purpose of the coordinated effort is to assist water suppliers to provide the highest possible quality of water economically and reliably. The true benefits are realized when the results are implemented at the utility level. The Foundation's trustees are pleased to offer this publication as a contribution toward that end.

Water treatment facilities are often faced with the challenge of trying to achieve conflicting treatment objectives in a single process train. In the treatment of surface water, it is not uncommon to find oxidant, coagulant, and disinfectant chemicals undergoing simultaneous reactions. The potential for chemical interactions to occur among water treatment chemicals is related to the relative reaction kinetics, the sequence of chemical addition, and water quality parameters. To optimize the effective use of chemicals in water treatment, it is important to consider potential interactions that occur. The focus of this project is to evaluate potential interactions of polyelectrolytes with other chemicals used for drinking water production.

Walter J. Bishop
Board of Trustees Chair
Awwa Research Foundation

James F. Manwaring, P.E.
Executive Director
Awwa Research Foundation

ACKNOWLEDGMENTS

The authors of this report are indebted to the following water utilities and individuals for their cooperation and participation in this project:

The utilities that participated in the study included:

- The City of Tampa Water Department, Tampa Florida.
- Metropolitan Water District of Salt Lake and Sandy, Sandy,Utah
- Champlain Water District, South Burlington, Vermont
- City of Fort Collins Water Utilities, Fort Collins, Colorado

The participating utilities provided access to data and information about their treatment systems, with an emphasis on chemical use information. They also provided samples of chemicals and source water. In addition, the City of Tampa Water Department provided assistance with analysis of disinfection byproducts. The involvement of the utilities in the project is highly appreciated.

The authors appreciate the support, input, and guidance provided by Traci Case, the Project Manager for this project. The Project Advisory Committee (PAC) for this project included Franklyn Smith, Regional Municipality of Waterloo Bill Becker, Hazen and Sawyer; David Smith, New York City Department of Environmental Protection; and Jan Routt, Kentucky American Water. The input, support and valuable suggestions of the PAC throughout the project are sincerely appreciated. They provided many practical recommendations that have been integrated into the final report..

The authors thank the utilities that participated in the on-line survey for the project and provided valuable data about water treatment chemicals and treatment practices. In addition, laboratory and computer assistance was provided at the University of South Florida by Joseph Viciere, Tianbo Xu, Maysoon Sallam, Xiongfei Xie, Tom Moran, Benjamin Jewell, Monica Wolfson, Paul Silver, Manimeldura De Silva, Qiushi Tang, Johna Jahn, Rochelle Minnis, and Peiyao Cheng. Filter precoat studies were conducted by N. J. Anderson, B. A. Bolto, D. R. Dixon, R. J. Eldridge and E. Rodda of CSIRO Molecular Science.

EXECUTIVE SUMMARY

Polyelectrolytes are used in water treatment systems for a variety of applications including coagulation, flocculation, ballasted sedimentation, filtration, dissolved air flotation, and sludge conditioning. There are multiple advantages to using polyelectrolytes including more efficient removal of particles, organics, and color; reduced sludge formation; sludge conditioning, and reduced reliance on metal salt coagulants. Effective use of polyelectrolytes can also yield cost savings. This project was developed to determine the degree to which organic polyelectrolytes react with other chemicals used in water treatment applications.

RESEARCH OBJECTIVES

The research objectives for this project are:

1. Evaluate current patterns of chemical use for drinking water production with respect to the types of chemicals used, chemical dosages, chemical addition practices, potential chemical interactions, and operational measures of chemical effectiveness.
2. Conduct bench-scale testing to evaluate polyelectrolyte interactions and the impacts of these interactions on process performance, water quality, and operations.
3. Assess the degree to which exposure to ultraviolet light impacts polyelectrolyte reactions with oxidants
4. Conduct a laboratory study to determine the effect of precoating filter media with various polyelectrolytes on filtration efficiency and filter backwashing effectiveness.

APPROACH

A key component of this project was to develop analytical methods to determine if residual polymers can be measured and if changes have occurred in the polymer structure that can influence its effectiveness, toxicity, or reactivity in downstream treatment processes. Direct and indirect methods were developed to characterize residual polyelectrolytes.

Indirect characterization methods included the development of screening tools to test polyelectrolyte reactivity to ultraviolet light and oxidants. Screening tools included organic carbon oxidation, absorption spectra, and reactions with oxidants. Other indirect characterization methods included charge density titrations, streaming current analysis, and viscosity measurements.

Performance measures for polymer effectiveness included tests of particle size, filtration rate, and residual turbidity. Disinfection byproduct formation potentials were also used to compare the relative degree of oxidation that occurs for individual polyelectrolytes. Specific performance parameters were developed for each treatment step.

Direct characterization methods included capillary electrophoresis and size exclusion chromatography. Detection limits of both techniques were too close to the dosage levels used in common practice in the water industry. A concentration step would be needed to allow for monitoring of residual polyelectrolytes by either of these methods. Other analytical tools that might be useful for quantifying residual polyelectrolytes include HPLC/MS or NMR. Gas chromatography can also be used to quantify monomers. These methods were not tested in this project.

Bench-scale and pilot-scale tests were conducted to determine the impacts of precoating filter media with different formulations of polyelectrolytes on filtration efficiency. The testing was conducted to determine if precoating filter media could increase filter run times and increase the rate of filter ripening. A fluorescently labeled version of one of the polyelectrolytes was used to determine the fate of the polyelectrolytes through water treatment systems. This approach could be useful for optimizing polyelectrolyte dosages and for identifying potential degradation products.

CONCLUSIONS

The major conclusions from this study are:

1. The dominant types of polyelectrolytes used in production of drinking water are polyacrylamides, polyDADMACs, and EPI-DMAs. Other chemicals that are used concurrently with polyelectrolytes include oxidants (permanganate, chlorine, ozone, chlorine dioxide), inorganic coagulants, lime, and corrosion inhibitors.
2. When polyelectrolytes are exposed to ultraviolet light in the presence of oxidant chemicals, advanced oxidation reactions can occur. These advanced oxidation reactions can cause decomposition of the polyelectrolyte resulting in reduced performance.
3. Evidence exists that some combinations of polyelectrolytes, chlorine, and ultraviolet light can result in the formation of disinfection byproducts. The levels of regulated disinfection byproducts (trihalomethanes and haloacetic acids) formed by polyelectrolyte reactions, when using typical polyelectrolyte dosages, can range from 4 to 47 μg DBP/mg polyelectrolyte.
4. The use of cationic or anionic polyelectrolytes of low charge density to precoat filter media can result in temporary increases of one to five hours in filter run times. Precoating filter media with a cationic polyelectrolyte of medium molecular weight and high charge density (1 mg/g) appeared to result in about a 20% decrease in the filter ripening period. However, the effects of precoating do not appear to be sustainable over more than three filter runs. Based on the quantities of polyelectrolytes needed for precoating the filter media and the lack of sustainability, this process does not appear to be cost-effective for conventional filter applications.
5. Fluorescent labels can be attached to polyelectrolytes to allow for the use of fluorescence monitoring. The preliminary tests conducted in this study suggest that further refinement of the fluorescent labeling technique could provide a useful tool for rapid monitoring of residual polyelectrolytes. The ability to monitor residual polyelectrolytes would help to improve polyelectrolyte dosage control and process performance.

RECOMMENDATIONS

The simultaneous use of multiple chemicals is widespread in the drinking water industry. While each chemical is employed for a specific purpose, all of the chemicals that are used have the potential to interact within the water matrix. The net effect of chemical interactions can be positive,negative, or neutral; depending on water quality. This research project has provided a framework for evaluating the extent of chemical interactions that occur during drinking water treatment.

Chemical costs represent a major operating expense for water utilities. Polyelectrolytes and oxidant chemicals are the some of the most costly chemicals. Through this project, it is clear that the concurrent presence of oxidant chemicals and polyelectrolytes can result in degradation of both types of chemicals, thereby decreasing process efficiency and introducing unnecessary operating costs.

Exposure to sunlight can introduce an additional oxidant demand on water and can also result in advanced oxidation reactions. When polyelectrolytes are added to water, either as part of treatment or through recycled flows, exposure to ultraviolet light can degrade the polyelectrolytes, compromising process performance. The advanced oxidation effects were observed to different degrees for polyacrylamides, polyDADMACs and EPI-DMAs, with the extent of reaction specific to the type of polyelectrolyte and water quality parameters.

There are five important recommendations to utilities that can be derived from this project.

1. **Use of polyelectrolytes in conjunction with other chemicals.**
   Polyelectrolytes are effective for improving the efficiency of coagulation, flocculation, ballasted sedimentation, and filtration. In the design and operation of chemical feed facilities, it is important that adequate reaction time be available to segregate polyelectrolyte addition from addition of other chemicals, particularly oxidant chemicals (i.e. permanganate, chlorine, ozone) to prevent carry-over of residual polyelectrolyte into downstream processes.
2. **Make-up water for polyelectrolyte addition**.
   Typically, polyelectrolytes are supplied in either dry form or as concentrated solutions. Prior to injection into the treatment train, dry chemicals are hydrated or dissolved. To optimize the effectiveness of polyelectrolytes, it is important that the make-up water is free of excessive levels of oxidants (i.e. chlorine residual or permanganate). Usually the most cost-effective make-up water is finished water, however, finished water frequently contains a disinfectant residual. It is suggested that preliminary tests be conducted to ensure the compatibility of the make-up water with the polylectrolyte solution.
3. **Indirect monitoring tools for polyelectrolyte application optimization**.
   Due to the relatively low concentrations of polyelectrolytes typically used in potable water treatment, direct monitoring of residuals is not practical. Indirect monitoring tools should be used to help ensure that polyelectrolyte dosing and mixing conditions are appropriate. Examples of indirect monitoring tools include streaming current, residual TOC, or TOC removal efficiency (particularly for waters with TOC levels below 2 mg/L), disinfection byproduct formation potential (and formation potential removal), oxidant demand, and UV_{220}.

   Routine monitoring should be conducted immediately after polyelectrolyte addition and downstream of all locations where polyelectrolytes are reacting and being removed (i.e. flocculation, sedimentation, or filtration). If polyelectrolytes are overdosed, or dosed too close to other chemical application points, potential reactions may occur with other chemicals, particularly in the presence of natural or applied UV light. Such reactions can result in the formation of disinfection byproducts or byproduct precursors and can also result in chemical fouling of filter media and UV lamps.

4. **Use of polyelectrolytes in reaction basins exposed to sunlight.**
Some types of polyelectrolytes decompose due to exposure to UV light. The decomposition reactions are exacerbated by the presence of chemical oxidants such as chlorine or permanganate. If possible, polyelectrolyte addition and mixing should be completed prior to sunlight exposure. Similar precautions should be exercised if polyelectrolytes are used upstream of UV disinfection processes.

5. **Recirculation of solids containing unreacted polyelectrolytes and reacted polyelectrolytes**.
In many water treatment facilities, residual solids are recirculated within the plant either from filter backwashing recycle streams, return flows from thickening basins, or via solids maintained intentionally in solids-contact upflow clarifiers. Residual solids may contain residual polyelectrolytes providing a mechanism for inadvertent introduction of polyelectrolytes back into the flowstream. When these concentrated zones of polyelectrolytes are exposed to other treatment chemicals and/or UV, the potential for chemical interactions can increase. It is suggested that steps be taken to avoid this problem in the development of dosing strategies for polyelectrolytes and other treatment chemicals. Design and operation of systems for solids removal and recirculation of return flows should also consider the potential impacts of polyelectrolyte interactions.

6. **Emerging health effects concerns**.
Currently, the Safe Drinking Water Act (SDWA) includes limits on the allowable concentrations of acrylamide and epichlorohydrin monomers in drinking water supplies. Recent research (AwwaRF 2678) and the genesis of revised AWWA standards have demonstrated that commonly used polyelectrolytes such as epichlorhydrin-dimethylamine (EPI-DMA) and poly(Diallydimethylammonium chloride (polyDADMAC) may react with disinfectants to form significant levels of N-nitrosodimethylamine (NDMA), a suspected carcinogen with a theoretical lifetime cancer risk level of 10^{-6} from exposures levels of 0.7 ng/L in drinking water (Valentine et al. 2004). This project helps to inform utilities about the potential for undesirable side-reactions associated with polyelectrolyte use and provides advice that can be applied towards design and operations concerns.

FUTURE RESEARCH

Because of the wide variety of potential chemical interactions that can occur in water treatment facilities, additional research is needed to assess the degree to which these reactions occur. The low levels of residual polyelectrolytes in treated water impose analytical challenges for process monitoring. Development of improved tools for quantifying residual polyelectrolytes, such as the use of fluorescent labels or antibody based analytical methods or sensors may be valuable in optimization of chemical dosages and reducing the potential for chemical interactions to occur. Research is also needed to quantify the potential impacts of residual polyelectrolytes on nontraditional technologies such as UV irradiation and membrane processes.

CHAPTER 1: INTRODUCTION

Organic polyelectrolytes have been used for drinking water applications since the 1960s (Bolto 1995, Cary et al. 1987, Crozes et al. 1995). Organic polyelectrolyte applications in drinking water include use as coagulant aids in conjunction with inorganic coagulants such asiron or aluminum salts (Ghosh et al. 1985, Kawamura 1976), as primary coagulants for turbidity removal (Amy and Chadik 1983, Bolto et al. 1998), as filtration aids (Edzwald et al. 1977, Yeh and Ghosh 1981, Zhu et al. 1996), and in sludge processing (Dentel et al. 1993). The effectiveness of specific polyelectrolytes for a given application is dependent on the treatment goals and water quality. Most polyelectrolytes used in water treatment alter the stability of particulate matter by adsorption and charge neutralization. When organic polymers are used as primary coagulants, the goal is to promote the growth of flocs by a combination of particle destabilization, charge neutralization, and precipitation. Some of the advantages of using polymers in water treatment include lower dose requirements of inorganic coagulants, a lower volume of sludge, a reduction in the ionic load of the treated water, reduced consumption of alkalinity, improved performance of filtration systems, and more efficient sludge dewatering and handling (Letterman and Sricharoenchaikit 1982, Rout et al. 1999; Nozaic et al. 2001).

POTENTIAL FOR POLYELECTROLYTE INTERACTIONS WITH OTHER CHEMICALS

The potential for chemical interactions to occur during water treatment depends on the sequence of chemical addition, water quality, and operational variables. In treatment of surface water, pre-oxidation is typically followed by addition of coagulant chemicals. For example, oxidant chemicals may be added to water intakes to treat reduced minerals (i.e. iron and manganese) and compounds associated with taste and odor. If the oxidation reaction is not completed during the pre-oxidation stage of the water treatment facility, residual oxidants may be available for further reactions as the water proceeds through the treatment plant.

Similarly, chlorine or other disinfectant chemicals may be added to water prior to sedimentation to allow for adequate contact time. Polyelectrolytes present in the water during flocculation and sedimentation could potentially react with disinfection chemicals. Other potential reactions include interactions with activated carbon (PAC or GAC), reactions with filter media surfaces, or downstream reactions with disinfectant chemicals, membranes, corrosion control chemicals, or fluoridation chemicals.

PROJECT OBJECTIVES

This project was developed to evaluate the degree to which chemical interactions occur during drinking water treatment with a specific emphasis on polyelectrolyte interactions. The project objectives are:

1. Evaluate current patterns of chemical use for drinking water production with respect to the types of chemicals used, chemical dosages, chemical addition practices, potential chemical interactions, and operational measures of chemical effectiveness.

2. Conduct bench-scale testing to evaluate polyelectrolyte interactions and the impacts of these interactions on process performance, water quality, and operations.
3. Assess the degree to which exposure to ultraviolet light impacts polyelectrolyte reactions with oxidants
4. Conduct a laboratory study to determine the effect of precoating filter media with various polyelectrolytes on filtration efficiency and filter backwashing effectiveness.

CHARACTERISTICS OF ORGANIC POLYELECTROLYTES USED IN WATER TREATMENT APPLICATIONS

Organic polyelectrolytes are used to facilitate removal of particulate material, natural organic matter (NOM), and to improve the settleability of flocs and the dewaterability of sludges. Polyelectrolytes used in drinking water treatment are classified in several ways: by net charge, molecular weight, charge density, and composition. In general, polyelectrolytes are classified as cationic, anionic, or nonionic (amphoteric) depending on the net charge on the molecule. Other important characteristics of polyelectrolytes include molecular weight and charge density. A summary of the characteristics of polyelectrolytes used in drinking water is given in Table 1.1. Specific applications, typical pre- and post- treatment requirements and dosage ranges are given in Table 1.2. Details on polyelectrolyte characteristics are provided in the Appendix. In addition to the applications listed in Table 1.2, polymers are also applied for sludge conditioning to facilitate dewatering. There is a potential for residual polymer in return flows from sludge dewatering to be introduced back into the treatment train.

Table 1-1
Summary of the types of organic polyelectrolytes used in drinking water treatment[1]

Chemical name	CAS number	Abbreviation	Dominant Charge	Charge density, mole percent	Relative Molecular weight	Monomers and structure	Form
Poly(diallyldimethyl-ammonium chloride)	26062-79-3	polyDADMAC	Cationic	10-25%	Low to medium	Allyl chloride and dimethylamine (DMA) Linear structure	Solution; total solids vary from 10-40% by weight
Epichlorohydrin dimethylamine polyamines	25988-97-0; 42751-79-1	EPI-DMA	Cationic	>25%	Low	Epichlorohydrin (EPI) and dimethylamine (DMA) Branched structure	Solution: total solids vary from 10-65% by weight
Polyacrylamides	9003-05-8	APAMs	Anionic	Wide range	High	Anionic co-polymers of acrylamide and anionic monomers of acrylic acid	Powders, emulsions, and solutions (high viscosity)
Polyacrylamides	9003-05-8	CPAMs or PAMs	Cationic or Nonionic	Wide range	High	Cationic copolymers of acrylamide and cationic monomers and nonionic homopolymers of acrylamide	Powders, emulsions, and solutions (high viscosity)

[1]Adapted from American Water Works Association, 1998, 1999, and 2001)

Table 1-2
Overview of the applications for organic polyelectrolytes used in drinking water treatment

Treatment application	Type of polymer	Chemical structure	Potential pre-treatment	Post-treatment	Dosage range, mg/L
Coagulant aid	Cationic or Nonionic	polyacrylamide polyDADMAC[1] EPI-DMA[2]	Oxidation PAC[3]	Flocculation Sedimentation Filtration Disinfection	0.2-2
Primary coagulant	Cationic	polyacrylamide polyDADMAC EPI-DMA	Oxidation PAC	Flocculation Sedimentation Filtration Disinfection	1-25
Flocculant aid	Cationic or Anionic	polyacrylamide polyDADMAC EPI-DMA	Oxidation PAC Coagulation	Sedimentation	0.2-2
Filter aid	Nonionic or Anionic	polyacrylamide	Oxidation Coagulation/ Flocculation Sedimentation Disinfection	Disinfection Membranes GAC[4]	0.01 to 2

[1]Poly(diallyldimethylammonium chloride); [2] Epichlorohydrin dimethylamine polyamines; [3]Powdered Activated Carbon; [4] Granular Activated Carbon

Molecular Weight

The molecular weight of organic polyelectrolytes depends on the monomer type, chain length, the presence or absence of co-monomers, and the degree of chain branching. Molecular weights of water treatment polyelectrolytes range from less than 10,000 to over 6,000,000 atomic mass units (amu). The molecular weight and physical structure can impact reaction rates (Lee and Gregory 1991) and the size and strength of flocs (Ghosh et al. 1985).

Charge Density

The charge density (CD) is an important property of polyelectrolytes. The CD reflects the net amount of charged sites that are available to react with constituents in water. The CD of polymers is expressed in terms of the percentage of functional groups that are charged (ionogenic) relative to all the functional groups in the polymer. The charge on some functional groups is independent of pH whereas other groups are only charged under specific conditions of pH, for example carboxylic groups are anionic under neutral or basic pH ranges and neutral under acidic ranges. The units used to express CD are either mole percent or meq/g, based on the chemical composition of the polymer.

The values for CD are not exact because, in many cases, polymers contain other constituents such as co-monomers, initiator fragments, or impurities thereby introducing sources of error into the calculation. The maximum CD for a given polymer formulation is highly variable and ranges from 5-15%, although levels as high as 45% have been reported (Wang and Audebert 1987; Eriksson et al. 1993). Charge densities of commonly used polyelectrolytes are shown in Table 1.3.

Table 1-3
Charge densities of commonly used polyelectrolytes

Charge classification	Polymer	Molecular Formula	CD, mol%	CD. meq/g
Cationic				
	EPI-DMA[1]	$C_5H_{12}ON\ Cl$	100	7.3
	PolyDADMAC[1]	$C_8H_{16}N\ Cl$	100	6.2
	CPAM[3]	$C_8H_{16}O_2N\ Cl$	100	5.2
	Chitosan	$C_6H_6O_4\ HCl$	100	5.2
	CPAM[3]	$(C_8H_{16}O_2N\ Cl)_{0.25}\ (C_3H_5ON)_{0.75}$	25	2.5
	CPAM[3]	$(C_8H_{16}O_2N\ Cl)_{0.1}\ (C_3H_5ON)_{0.9}$	10	1.2
Anionic				
	APAM[4]	$C_3H_3O_2\ Na$	100	10.6
		$(C_3H_3O_2\ Na)_{0.75}\ (C_3H_5ON)_{0.25}$	75	8.5
		$(C_3H_3O_2\ Na)_{0.50}\ (C_3H_5ON)_{0.50}$	50	6.1
		$(C_3H_3O_2\ Na)_{0.25}\ (C_3H_5ON)_{0.75}$	25	3.3
		$(C_3H_3O_2\ Na)_{0.1}\ (C_3H_5ON)_{0.9}$	10	1.4

[1]Epichlorohydrin/dimethylamine [2]Poly(diallyldimethyl ammonium chloride) [3]Cationic Polyacrylamide; [4]Anionic polyacrylamide

WATER QUALITY VARIABLES RELEVANT TO THE USE OF ORGANIC POLYELECTROLYTES

The use of polyelectrolytes for drinking water applications is based on modifying the properties of colloidal material and/or dissolved organic matter to improve removal by liquid/solid separation processes such as sedimentation and filtration. Water quality factors that are important in the use of organic polyelectrolytes include the nature of the particulate matter and dissolved organic carbon, pH, and ionic strength.

Colliodal and Suspended Particles

Particulate matter in water ranges in size from less than 0.1 to over 10 μm and consists of insoluble mineral precipitates, debris, clay, and microorganisms. Inorganic particles in natural waters include silica, aluminosilicates, iron, and manganese oxides. The amount of particulate matter in untreated water can vary from 1 to 200 mg/L depending on the water source and seasonal variations (i.e. rainfall and runoff). Removal of particulate matter from water is a major focus of surface water treatment facilities to allow for effective disinfection and to prevent shielding of microorganisms within distribution systems. Polyelectrolytes are used to facilitate removal of particulates by modifying the surface characteristics through coagulation and adsorption prior to solid/liquid separation (i.e. sedimentation or filtration).

In addition to particle size, surface charge is an important parameter influencing particle removal. In natural waters, the net surface charge on particles tends to be negative and can be monitored via particle microelectrophoresis (Beckett and Le 1990). The charge on particle surfaces is controlled by an adsorbed layer of natural organic matter (NOM), conductivity, and the concentrations of divalent cations in the water (Beckett 1990). The adsorption of organics on particulate matter has a strong impact on the net surface charge and can therefore influence the amount of coagulant required and the rate of coagulation (Gibbs 1983).

Natural Organic Matter (NOM)

Natural organic matter (NOM) consists of macromolecular and particulate organic material that is stable in water. NOM imparts color, taste, and odor to water and serves as a precursor to the formation of disinfection by-products (DBPs). In addition, NOM interferes with the oxidation of dissolved iron and manganese, fouls membranes and ion-exchange resins, and can provide a substrate for the proliferation of biofilms in distribution systems (Huang and Yeh 1993). The levels of NOM in untreated surface waters vary with watershed characteristics and are reported as total organic carbon (TOC) (Boggs et al. 1985). Typically, TOC concentrations range from about 2 mg/L for an oligotrophic lake, 5 mg/L for a river, 10 mg/L for a eutrophic lake, 15 mg/L for a marsh to over 30 mg/L for bog water (Thurman 1985). Groundwater can contain TOC ranging from 0.1 to 10 mg/L.

The characteristics of NOM important for polyelectrolyte reactions include the chemical composition, the functional groups, and the molecular size. NOM can contain amino acids, fatty acids, phenols, sterols, sugars, hydrocarbons, urea, porphyrins and polymers such as polypeptides, lipids, polysaccharides, and polydisperse humic substances with molecular weights ranging from several hundred to over 10,000 amu. In natural waters, humic substances tend to be acidic and predominantly aromatic negatively charged compounds with surface active properties. These variable properties influence reactivity (Stumm 1992). Humic substances can be subdivided into humic acids, which are soluble in alkali but not acid; and fulvic acids, which are soluble under acidic and alkaline conditions; and humans, which are not soluble in either acid or alkali. The proportion of humic materials in surface waters ranges from about 30 to 80% of the NOM (Lawrence 1989).

In comparison to humic acids, fulvic acids are lower molecular weight hydrophilic compounds and tend to contain twice as many carboxylic acid groups, 10% more oxygen, and about 10% less carbon. The amount of alcohol groups is similar. The molecular weights of fulvic acids range from 300 to 3000; whereas the molecular weight range for humic acids is from about 700 to approximately 10,000 (Thurman 1985). The functional groups are predominantly phenolic and carboxylic acids, with some alcohol, quinone, ether, ester and ketone groups. A comparison of the dominant functional groups of fulvic and humic acids is shown in Table 1.4 (Boggs et al., 1985).

Table 1-4
Comparison of typical groups in fulvic and humic acids[1]

Property	Fulvic acids, meq/g	Humic acids, meq/g
Total acids	10.3	6.7
RCOOH	8.2	3.6
Phenolic OH	3.0	3.9
Aliphatic OH	6.1	2.6
Quinone, ketone	2.7	2.9
OCH3	0.8	0.6

[1] Adapted from Boggs et al. 1985

Humic substances contain hydrophobic aromatic and aliphatic regions. Because humic acids are less acidic than fulvic acids, the conjugate anions are stronger bases and metal ions are bound more strongly by humic acids than fulvic acids (Tipping 1993). In coagulation with metal salts, humates bind more strongly to the metal than fulvates. Therefore, waters containing large amounts of fulvic substances are more difficult to treat by coagulation processes, especially if turbidity levels are low (Edzwald 1993). The lower molecular weight impurities are better removed by adsorbents (Johannsen et al. 1993), ion exchange (Bursill et al. 1985), oxidation, or biological methods (Stumm 1992).

pH

pH can impact the surface charge on particulate matter and NOM In addition, hydrolysis of some types of organic polyelectrolytes is influenced by pH, particularly where cationic functional groups are attached to the polymer backbone by ester linkages. The type of functional groups and the CD influence the stability and half life of a given polymer. For a polymer with a CD of 24 mol %, the half life decreased from 24 hours at pH 7 to 15 min at pH 8.5 (Aksberg and Wagberg 1989). Less ionic polymers with CDs of about 6% had half lives of 22 months at neutral pH (Smith-Palmer et al. 1994). In addition to the loss of reactive sites, hydrolysis causes a change in the chain conformation due to the formation of anionic carboxylates, which reduces the chain extension and makes the polymer less efficient as a flocculant. In the case of homopolymers the corresponding methacrylate is less vulnerable to hydrolysis (van de Wetering et al. 1998).

Ionic Strength

Ionic strength can impact the effectiveness of organic polyelectrolytes in water. A comparison of the influence of ionic strength on charge density is shown in Table 1.5. For low CD polyelectrolytes, competitive forces between the polymer and the particle surface are increased by increasing ionic strength. Conversely, for high CD polyelectrolytes, increases in ionic strength tend to improve adsorption by reducing the electrostatic repulsive forces between the polymer segments (van de Steeg et al. 1992).

Table 1-5
Comparison of Ionic Strength Impacts on the Performance of Organic Polyelectrolytes of Varying Levels of Charge Density

Charge Density	Range	Impact of ionic strength on reaction with clays
Low	1%	Decreased adsorption
Intermediate	5%	No impact
High	13-30%	Increased adsorption

WATER TREATMENT APPLICATIONS

Organic polyelectrolytes are used for coagulation, flocculation, sedimentation, filtration, dissolved air flotation, ballasted sedimentation, and sludge conditioning. A review of these applications is presented in this section.

Coagulation/Flocculation

Coagulation is the destabilization of particulate and/or macromolecular impurities due to chemical addition. Flocculation is the process of aggregating the destabilized particles to form larger agglomerates or flocs (Hutchison and Healy 1990). To destabilize negatively charged impurities in water, positively charged species are added in appropriate quantities to neutralize the charge. A flocculation step is then used to aggregate the particles to produce larger flocs. Organic polyelectrolytes have been applied for coagulation/flocculation processes for water purification for over four decades (Kawamura 1976).

The use of hydrolysable metal salts for coagulation is widely practiced (Stumm 1992; Rebhun and Lurie 1993). Polyelectrolytes can be used in tandem with metal salts or as primary coagulants. One of the benefits of using polyelectrolytes is to increase the rate of particle sedimentation in cold waters by forming larger and more dense flocs. In addition, polyelectrolytes can be effective for improved floc formation in soft waters (Faust and Aly 1983). In some cases, the use of polyelectrolytes to produce larger and stronger flocs can result in more effective sedimentation thereby increasing the capacity and efficiency of a treatment facility.

The best turbidity and color removal occurs when the mobility or zeta potential of the particles is close to zero (Edzwald, 1986). Waters of low turbidity and moderate color are ideal candidates for polymer coagulation. Converesely, the use of polyelectrolytes as the main coagulant for waters of high color and high TOC can result in increased formation of disinfection byproducts. In general, the use of polyelectrolytes can act to reduce the amount of inorganic coagulants needed for coagulation, resulting in the production of less sludge (Schlauch 1981; Vik and Eikebrokk 1989). However, precise control of dosages is difficult and it is possible that residual polymers or monomers can carry over into the product water. The level of residual polymer is minimized when the turbidity is high (Bennett et al. 2000).

Mechanism of action

Several concurrent reactions occur during coagulation/flocculation processes. The dominant mechanisms of action include charge neutralization, electrostatic charge patch destabilization, and polymer bridging.

Charge Neutralization. For destabilization of negatively charged particles, the dominant mechanism of action is charge neutralization. When hydrolyzable metal salts such as aluminum or iron are used as primary coagulants, soluble hydrolyzed cations and metal hydroxide flocs are formed, with the relative amounts related to the operating pH. For example, alum can form monomeric hydroxy species like $AlOH^{2+}$, $Al(OH)_2^+$ and $Al(OH)_4^-$ within about 10^{-4} seconds. After about 1 second, concurrent formation of insoluble $Al(OH)_3$ precipitates, $Al_2(OH)_{24}^+$ dimers, and metallic polymers occurs. The metallic polymers can contain up to thirteen aluminum atoms and five positive charges. Several pH dependent mechanisms have been postulated for the coagulation of NOM and turbidity particles by aluminum ions. Under low pH conditions (<5) organic acids are at least 50% ionized and aluminum humates are formed:

$$2\,RCOO^- \quad + \quad AlOH^{2+} \quad \rightarrow \quad (RCOO)_2AlOH \qquad (1)$$

At pH levels between 5 and 6, polynuclear aluminum species react with NOM and particulate matter. For pH levels above 7, adsorption of the organics on aluminum hydroxide occurs by the ligand exchange reaction:

$$RCOO^- \quad + \quad OH\text{-}Al< \quad \rightarrow \quad RCOO\text{-}Al< \quad + \quad OH^- \qquad (2)$$

Analogous reactions occur with aluminum humates formed under low pH conditions from mono or polynuclear hydroxy forms of aluminum. The flocs consist of a mixture of aluminum hydroxide precipitates, NOM, and turbidity particles. Alternatively, inorganic polymeric coagulants (e.g. poly(aluminum chloride) or poly(aluminum sulfate)) can be applied as coagulants with optimum dosages up to 30% lower than metal salts (Nilsson and Kvant 1986). The inorganic polymeric species have higher charge densities that function to neutralize negative charges on suspended clay particles and NOM (Singer and Harrington 1993). The polymeric forms are more effective than the metal salts for treatment of cold waters and can be employed over a wider pH range, but tend to have higher costs. Generally, alum is reported to be more effective for removal of NOM and turbidity than aluminum polymers for average waters, whereas iron polymers tend to outperform the monomeric species (Jiang et al. 1993).

The dominant use of organic polyelectrolytes in water treatment is as a coagulant aid to bridge the coagulated particles formed when aluminum or iron salts have been used as the primary coagulant. An appropriate polyelectrolyte can increase floc size and density enabling more rapid sedimentation. When organic polylectrolytes are used in conjunction with inorganic coagulants, the optimum dose and type of polymer depends on the net surface charge of the flocs formed through coagulation. Typically, the net charge on flocs formed by inorganic coagulants is either slightly positive or slightly negative, depending on the effective dose of inorganic coagulant at the floc surface. For positively charged flocs, an anionic polyacrylamide of high molecular weight and low or medium CD is appropriate. When the coagulated solids have a slightly negative charge, a cationic polymer of high molecular weight and low CD, such as a cationic polyacrylamide, is effective in forming larger flocs (Edzwald 1977, 1986). The use of polymers as coagulant aids can result in a 40 to 60% reduction in the dose of inorganic coagulants (Hall and

Hyde 1992, Lee and Gregory 1991). However, it is important to note that an excess of polymer can cause re-dispersion of the impurities.

When organic polyelectrolytes are used as primary coagulants, the polymer plays a dual role as both a destabilizing agent and an agent for floc growth. Destabilization occurs via a charge neutralization/precipitation mechanism. At optimum polymer dosages, the surface charge is nearly eliminated so that only weak electrostatic repulsion remains. Charge neutralization can occur under conditions of either low or high surface coverage, depending on the type and density of the charges associated with the polymer and the particle surfaces (Hongeveen et al. 1996). The characteristics of organic polyelectrolytes that are important for charge neutralization include low to medium molecular weight and a high CD. The best turbidity and color removal occurs when the electrophoretic mobility or zeta potential of the particles is close to zero. The CD rather than the molecular weight is important in selecting the optimal conditions, and the initial periods of both rapid and slow mixing are critical in the formation of flocs (Leu and Ghosh 1988). It has been reported that longer chained polyelectrolytes are more effective for treatment of waters with higher turbidity (Lee et al. 2001).

Electrostatic charge patch destabilization. Under low polymer dosages, the mechanism of action has been described as electrostatic charge patch destabilization (Gregory 1985a, Norman and Seddon 1991). The polymer adsorbs at locations of opposite charge leading to excess charge on the polymer surface and regions of variable charge on the floc. Direct electrostatic attraction between the regions of opposite charge in adjacent particles results in flocculation. When a cationic polymer adsorbs on a negatively charged surface, patches of positive surface are formed which may attach to negative patches on another particle. Patch formation occurs when the surface is not completely covered, as in the initial stages of adsorption or with a low polymer dose. Because patching occurs immediately after addition of the polymer, it is sensitive to mixing conditions and can be characterized as a flocculation phenomenon. A characteristic of the electrostatic patch model is that flocs may reform after floc breakage. Polymer characteristics favoring the patch charge model are low dosages of high molecular weight polymer with high CD under low ionic strength conditions. For example, highly charged polymers adsorb strongly onto silicate mineral particles and silica resulting in low surface coverage (Parazak et al. 1988).

Polymer Bridging. Organic polyelectrolytes can be used to aggregate small flocs formed by coagulation with metal salts producing particles with more rapid rates of sedimentation. This process involves polymer bridging, in which looped and dangling chains of polyelectrolyte bound to a floc particle can attach to nearby particles. This mechanism is favored when long chain polymers that do not have a high level of charge are used. In bridging, one polymer chain adsorbs on two or more particles simultaneously, via an electrostatic or non-electrostatic driving force (Gregory 1985a). It takes place when the surface is not completely covered, as in the initial stages of adsorption or when the polymer dose is low. Because bridging occurs immediately after addition of the polymer, appropriate mixing conditions are critical (Hongeveen et al. 1996). Bridging is strictly a flocculation phenomenon, as the polymer is directly involved in linking the particles together.

Polymer characteristics that promote bridging include high molecular weight and low CD. Bridging requires a minimum molecular weight of about 800,000 amu with higher molecular weight polymers resulting in better performance (Wang and Audebert, 1987; Eriksson et al. 1993). Higher CD polymers tend to improve shear resistance, the transition occurring at a CD between 10 and 20%.

Multiple reactions. A summary of the properties of polyelectrolytes important for coagulation/flocculation mechanisms is shown in Table 1.6. In general, there is a balance between charge neutralization with highly charged polymers and inter-particle bridging when the polymer has a low content of ionic groups (Bolto 1985, Durand-Piana et al. 1987).

Table 1-6
Comparison of properties of organic polyelectrolytes important for coagulation/flocculation.

Mechanism of Action	Molecular Weight Range	Charge Density
Charge Neutralization	Low	High
Charge Patch	High	High
Interparticle Bridging	High (> 800,000 amu)	Low (<10%)

Filtration

Filtration is a critical step for removal of impurities from water. In conventional water treatment, filtration is used to remove particles that are too small to be removed by gravity sedimentation. Alternatively, for low turbidity waters (< 5NTU), direct filtration is used (Vigneswaran and Ben Aim 1989). Typically organic polyelectrolyte dosages range from 0.5 to 5 mg/L. Chemicals are mixed with water immediately prior to filtration to produce small pin-sized flocs that enter the body of the filter and adsorb to the filter media (Edzwald 1986, Reuter and Landscheidt 1988, Jackson 1980).

The use of cationic polyelectrolytes for direct filtration has advantages over the use of inorganic coagulants because charge neutralization can occur without the formation of additional solids. In contrast, the flocs from inorganic coagulants tend to accumulate in the filter media resulting in rapid increases in headloss and the need for frequent backwashing (Coccagna, 1989). The initial mixing intensity and the mixing time are the most important parameters in determining the particle size of the floc formed prior to filtration (Jackson, 1980).

The capacity of a filter plant that has been operated with inorganic coagulant may be increased by polyelectrolyte addition. For example, adding 0.3 mg/L of cationic polymer after ferric chloride resulted in lower dose requirements for ferric chloride, increased filter cycle times, increased capacity, and an improvement in the product water quality (Reuter and Landscheidt 1988). Zhu et al. (1996) investigated the effect of treated lime-softened water using a range of polyelectrolyte types and charges. The use of low dosages of polyelectrolytes (0.01 and 0.02 mg/L) resulted in longer filter runs and lower product water turbidity. For filtration of waters high in organics, a substantial polymer dose is needed, but compared to alum usage the filter run can be more than double in length (Edzwald 1986). For treatment of humic substances conventional metal ion coagulants can be used with cationic polymers, resulting in lower dosages of inorganic coagulants. High CD cationic polymers in the charge neutralization mode are widely used for filtration applications (Rebhun et al. 1984).

Cationic polymers can coagulate freshwater algae at doses of 1 to 10 mg/L (Bilanovich et al. 1988). For direct filtration, high CD cationic polymers can be effective if the doses are adequate to overcome the effects of exocellular matter released by the algae (Haarhoff and Cleasby 1989). Employing an inorganic coagulant for algae removal can disrupt the cell membrane and potentially cause release of algal exudates such as taste and odor compounds or cyanobacterial toxins (Kenefick et al. 1993).

Another approach for using polyelectrolytes in filtration is to coat the surfaces of the filter media, thereby reversing the surface charge. Baumann and Oulman (1970) found that using cationic polyelectrolytes to coat the surface of Celite fitler aids (0.4 mg polymer/g filter aid) led to enhanced adsorption of negatively charged impurities. Chaudhuri et al. (1974) investigated the use of uncoated and polyelectrolyte coated filter aid for removing T4 and MS2 viruses by filtration. They found that improved removal was associated with the coated filter aid, however the process efficiency was influenced by pH, filtration rate, and the presence of organic matter and turbidity. Brown et al. (1974) found that virus removal by diatomaceous earth could be increased substantially by either coating the filter material with cationic polyelectrolyte at a level of 0.5-1.0 mg polymer/g diatomaceous earth (DE), or by pre-addition of the polyelectrolyte to the water containing the virus prior to filtration. Amirhor and Engelbrecht (1975) confirmed the removal of MS2 virus on diatomaceous earth that had been coated with a polyamine of 60,000 MW at a level of 0.2-0.4 mg/g DE. They also found that coating improved removal.

Dissolved Air Flotation

In dissolved air flotation, a proportion of the treated water is recycled through a pressurized air saturation system, and the air-saturated water then released into the water to be treated (Hall and Hyde, 1992). Suitable polymers can prevent floc shearing brought about by overly vigorous contact of air bubbles with the particles (Walzer, 1981). In other work on dissolved air flotation, a treated water turbidity of 1 NTU was obtained when the original backwash water turbidity was in excess of 50 NTU (Eades et al., 2001). Chemical requirements consisted of a single site-specific polymer (0.1-0.5 mg/L) to form agglomerates that were suitable for flotation.

Ballasted Sedimentation

Ballasted sedimentation involves the addition of an inert ballasting agent such as silica sand or chemically conditioned sludge to a partially flocculated suspension (Metcalf and Eddy 2003). Organic polyelectrolytes play a major role in ballasted sedimentation. Typically, inorganic coagulants are coupled with cationic polyelectrolytes and mixed with the ballasting agent to facilitate binding of the floc particles.

Treatment of Filter Backwash Recycle

It is fairly common practice in the water treatment industry to conserve water by recycling spent filter backwash water, however, it is important to prevent recycling of contaminants through return flows. For example, it has been reported that protozoan pathogens can be concentrated up to 100 times in the backwash water as compared to the original untreated source water (Colbourne 1989, Cornwall et al. 2001 LeChavellier et al. 1991). Similar trends have been observed for concentration of NOM and other contaminants.

In a survey of the 335 water treatment plants in the US that recycle spent filter backwash water, the average generation of spent waters was found to be 2.5% (Arora et al. 2001). Most of the plants (88%) use surface water as their source, with 83% of these recycling to the head of the plant and only 2% to just before the filters. About 65% of the plants treat the filter backwash water before recycling by a variety of approaches including: sedimentation with or without added coagulants, dissolved air flotation with and without polymer, oxidation, conventional filtration or membrane filtration. Pilot studies of various methods to establish appropriate strategies showed that 0.5 mg/L of a cationic polyacrylamide of very high MW and medium CD removed 99.6% of the turbidity after filtration, and a similar dose of an anionic polyacrylamide removed 99.4% (Arora et al. 2001; Cornwall et al. 2001). The polymers were selected as the best options based on comparing results from jar tests on two cationic, three anionic and three non-ionic polymers. In a full-scale test, addition of only 0.1 mg/L of polymer to a side-stream lowered the settled turbidity by 50% (Cornwall et al. 2001).

One disadvantage of recycling filter backwash water treated with polyelectrolytes is that residual polymer may carry over into the recycled flow streams. Because recycle streams may be mixed with either untreated water or pre-oxidized water, there is a potential for residual polymers to contact the water prior to the addition of other water treatment chemicals. In a turbid water, some floc building may occur, potentially affect the nature of the final flocs and impacting settling properties.

Sludge Thickening

Sludge thickening is used to reduce the volume of sludge for disposal (Anderson et al. 1993a; Petruzzelli et al. 2000). Polyelectrolytes are used during thickening to produce large dense flocs with higher settling velocities. Typically, the supernatants from thickening are recycled. Polymers can also improve dewatering characteristics during centrifugation and filtration. The various methods used for concentrating water treatment plant sludges are summarized in Table 1.7 (Hall and Hyde 1992). Typically, in centrifugation 1.5-3 kg of polymer is used per ton of dry solids (Alt 1989). The high shear forces present during centrifugation make the use of polymers essential.

Table 1-7

Concentration of sludges from water treatment plants[1]

Conditioning method	Concentration in, % solids	Concentration out, % solids
Batch sedimentation	0.03-0.2	1-3
Continuous thickening		
without polymer dosing	0.03-0.2	2-3
with polymer dosing	0.03-0.2	5-10
Centrifugation	1-5	12-17
Filter press	1-10	20-25

[1]Adapted from Hall and Hyde 1992

For sludge treatment, polymers of high molecular weight and low or medium CD tend to give the best performance, with the charge type depending on the origin of the sludge. Thus for a water treatment plant sludge where alum is the coagulant, the sludge particles generally have a positive surface charge, so an anionic polymer is appropriate, although neutral polymers such as polyacrylamide are also employed. Such an alum sludge in a typical gravity filter operation at a solids loading of 25 $kg/m^2/day$ would have 1.5-2 % solids in the underflow without polymer use, and 3-4 % solids when an anionic polyacrylamide is present (Schlauch 1981). In comparison, biologically produced wastewater sludges tend to have negative surface charges, so cationic polymers tend to be more effective.

The mixing or pumping of sludges can impact dewatering because of shear stresses. When a high molecular weight medium CD anionic polyacrylamide is used for dewatering alum sludge, the dewatering response is dependent on both the shear and the mixing time (Novak and Bandak 1994). The polymer requirements increase with longer mixing times and at higher shear because of the disaggregation of the sludge. A higher polymer dose is then needed to re-agglomerate the particles created by excessive mixing conditions.

The influence of the CD of cationic polymers on biosolids has been explored (Tiravanti et al. 1985; Erikkson and Alm 1993). A low CD produces flocs with a flexible structure and good shear resistance, but the flocs are not exceptional in their filtration properties. A medium or high CD on the other hand results in good filtration properties and higher solids content, but the flocs are more sensitive to shear. A dual polymer method involving a high CD, low MW cationic polymer and, after suitable aging, a high CD, high MW cationic polymer has been successfully used to dewater a biological sludge, to give a final solids content of 11-14% versus the 8-11% obtained by conventional methods (Senthilnathan and Sigler 1993).

A recent development has been the synthesis of high MW cationic polyacrylamides that have been crosslinked (Mohammed et al. 2000). Crosslinked polymers are much less deformable and therefore cannot spread over the surface of the particle, preventing charge neutralization. Some charge remains for flocculation or re-flocculation, to give the polymer unique dewatering characteristics. In sludge conditioning, the massive agglomeration that takes place is initially capable of some breakdown with subsequent shear. Ultimately equilibrium is reached and the floc structure assumes a stable state. The floc size can be 30% larger with a crosslinked polymer than that obtained with the corresponding linear polymer, and the reduction in size on applying shear about 10% instead of more than 90%. Re-flocculation can take place producing a very stable floc. Crosslinked polymers have been most successful in centrifugation dewatering of sludges, when higher solids content corresponds to significant cost savings. Other applications include dewatering via belt presses and the sedimentation of slurries subject to high shear. Another advantage of these materials is that overdosing does not cause re-stabilization of the slurry, since adsorption of the polymer onto the surface of a particle cannot result in it occupying all of the active sites. The outcome is that there are higher throughput rates, a higher solids content of centrifuge cakes, and cleaner supernatants. Practical difficulties can arise in converting an alum treatment system to polylectrolyte coagulation (Nozaic et al. 2001). The clarifier hoppers designed for alum tend to have side wall angles from 40 to 50°, which are tend to result in the polymer flocs adhering to the walls instead of falling into the hopper.

TYPES OF POLYMERS AVAILABLE FOR USE IN WATER TREATMENT APPLICATIONS

Organic polyelectrolytes are classified as cationic, anionic, or nonionic. Many reviews of polymeric flocculants have been published covering synthesis, applications and the range of commercial organic flocculants available (Hoover 1970; Gutcho 1977; Vostrcil and Juracka 1976; Schwoyer 1981; Luttinger 1981). A summary of the major types of polyelectrolytes used in the water treatment industry is given here. Details of commercially available polyelectrolytes are given in the Appendix.

Cationic Polymers

The types of cationic polyelectrolytes that are used in water treatment include Poly(diallyldimethyl ammonium chloride) (PolyDADMAC); Epichlorohydrin/dimethylamine (EPI-DMA), cationic polyacrylamides (CPAM), and natural materials such as chitosan or starch.

Poly(diallyldimethyl ammonium chloride)

Poly(diallyldimethyl ammonium chloride) or PolyDADMAC is widely used for coagulation/flocculation in the water industry. It is produced from polymerization of diallyldimethylammonium chloride. PolyDADMAC is a water soluble polymer of low to medium molecular weight and contains five-membered pyrrolidinium units (Lancaster et al. 1976) as shown in Figure 1.1. Higher molecular weight copolymers have been produced by combining PolyDADMAC with acrylamide (Bolto 1995; Wandrey and Jaeger 1984). The copolymer is typically used for sludge conditioning.

Figure 0-1 Structure of PolyDADMAC. (Source: Reprinted from *Progress in Polymer Science*, Vol. 20, Bolto, B. A., Soluble polymers in water purification. pp. 987-1041, Copyright 1995 with permission from Elsevier.

Epichlorohydrin/dimethylamine polymers EPI-DMA

Epichlorohydrin forms polymers with ammonia and primary and secondary amines (Vorchheimer, 1981). The reaction of epichlorohydrin with a secondary amine such as dimethylamine produces the low molecular weight linear polymer epichlorohydrin/dimethylamine or EPI-DMA. In EPI-DMA, all of the nitrogens are in the quaternary ammonium form. The term 'polyamines' is sometimes misused to describe these polymers.

Epichlorhydrin, one of the monomers of EPI-DMA is regulated in drinking water under the Safe Drinking Water Act (SDWA) because it is considered to have the potential to function as a hepatotoxin, renal toxin, impact the central nervous system, or cause skin irritation due to short-term exposure (USEPA, 2002). Chronic exposure can potentially result in chromosome aberrations. Because it is a suspected carcinogen, the Maximum Contaminant Level Goal (MCLG) for epichlorohydrin is zero mg/L, and the maximum level derived from polyelectrolytes must be below 0.01% when dosed at 20 mg/L (USEPA 2002). The MCL is based on treatment technique (TT) by limiting the dose of EPI-DMA polyelectrolytes (see Appendix).

Cationic Polyacrylamides (CPAM)

Cationic polyacrylamides or CPAMs are formed from random copolymers of acrylamide and the cationic acrylate acryloyloxyethyltrimethylammonium chloride. Quaternization of dimethylaminoethyl acrylate with methyl chloride is used to form acryloyoxyehtyltirmethylammonium. CPAMs are extensively used in the water industry (Hoover 1970; Baade et al. 1989). The methacrylate analogue is also commercially available. The cationic content in the cationic polyacrylamides or CPAMs can be 10-100%. The structure of CPAMs is shown in Figure 1.2.

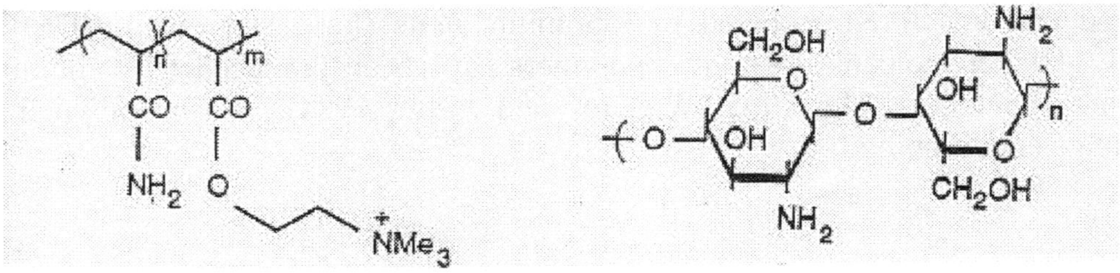

Figure 0-2 Structure of Cationic polyacrylamides (CPAMs) (Source: Reprinted from *Progress in Polymer Science*, Vol. 20, Bolto, B. A., Soluble polymers in water purification. pp. 987-1041, Copyright 1995 with permission from Elsevier.

Acrylamide, the polyacrylamide monomer, is regulated in drinking water under the SDWA because it is considered to have the potential cause damage to the central nervous system, paralysis and it is a suspected carcinogen (USEPA, 2004). The MCLG is zero mg/L, and the maximum acrylamide level derived from polyacrlyamides must be below 0.5 μg/L (USEPA 2004). The MCL is based on treatment technique (TT) by limiting the dose of polyacrylamides (see Appendix).

In copolymers with acrylamide, hydrolysis of the ester groups and consequent loss of cationic charge has been investigated and found to be CD and pH dependent, with hydrolysis increasing under more alkaline conditions:

$$\text{Pol-COOCH}_2\text{CH}_2\text{NMe}_3^+ + \text{OH}^- \rightarrow \text{Pol-COO}^- + \text{HO-CH}_2\text{CH}_2\text{Nme}_3^+ \quad (4)$$

By reacting polyacrylamide with formaldehyde and a secondary amine such as dimethylamine, followed by quaternization, a cationic polymer is formed that has a $–CH_2Nme_3^+$ group attached to the nitrogen of the amide group. (Vorchheimer, 1981).

Polydimethylaminoethyl methacrylate (PDMAEMA), is a homopolymer of moderate molecular weight (~1,000,000), that comprises a weakly basic variation of the cationic moiety in CPAM, but has a methacrylate rather than acrylate ester structure and a tertiary amino rather than a quaternary ammonium cationic group. Although the CD decreases with increasing pH, the polymer is fully protonated at pH levels below 6, and is 97% charged even at pH 7, so under normal conditions the CD of the homopolymer is very high (Hongeveen et al. 1996). Hydrolysis occurs at elevated pH levels, but is less pronounced than for the corresponding acrylate (van de Wetering et al. 1998). PDMAEMA is not used in potable water treatment applications.

Polyethyleneimine (PEI) is a low MW (usually about 10,000) weakly basic polymer which contains a mixture of primary, secondary and tertiary amino groups, generally in the ratio of 1 : 2 : 1 (Vorchheimer, 1981). Like PDMAEMA, its CD is pH dependent, but PEI is a much weaker base, with a pKa of 5.8 versus 8.6 for PDMAEMA. Consequently it has about 40% of the amino groups protonated at pH 6, and only 6% at pH 7 (Bloys Van Treslong and Staverman 1974).

Natural Cationic Polymers

There are several naturally occurring polymers that have inherent cationic properties or can be modified to yield a cationic polyelectrolyte (Bolto 1995). The most prominent of these is chitosan, a partially deacetylated chitin obtained from crustacean exo-skeletons that can be considered as a 1:4 random copolymer of N-acetyl-α-D-glucosamine and α-D-glucosamine (IV). It has also been derived from distillery wastewaters using a fungus-based process (Yokoi et al. 1998). The commercial product is of medium molecular weight and has a CD that is pH dependent and can be as high as 80% (Muzzarelli 1977). The optimal conditions for preparing chitosan for coagulation applications have been studied (Huang et al. 2000). Its use in water purification has been extensively reviewed (Pariser and Lombardi 1989). The adsorption of humic acid by both chitosan and chitin has received attention (Wan Ngah and Musa 1998). A quarternized version of chitosan is reported to have antibacterial activity (Kim et al. 1997).

Starch, a polymer composed of α-D-glucose units, can be converted to a cationic polymer with the cationic site attached via an ether link to the polymer chain (Harrington and Engelhardt 1979). The product has medium MW and the CD is low. A number of flocculating agents have been reported which are lignin based, being prepared by modifying kraft lignin to give it cationic character (McKague 1974).

Anionic Polyelectrolytes

The most common anionic polymers are those containing weakly acidic carboxylic acid groups, where the CD is dependent on pH. Polymers containing sulfonic acid groups that are fully ionized at all pH levels are employed in scale control. Anionic polyelectrolytes include anionic polyacrylamides and naturally occurring polymers such as lignin sulphonate.

Anionic Polyacrylamides

High molecular weight, low CD carboxylic acid polymers based on polyacrylamide are used extensively as flocculating agents in water treatment. Copolymers can be prepared either by copolymerization of acrylamide and acrylic acid or its salts, or by polymerization of acrylamide followed by partial hydrolysis (Vorchheimer 1981). The former route gives a roughly random copolymer, whereas some clustering of anionic groups can occur under alkaline hydrolysis. The CD can be determined by potentiometric titration of the copolymers.

Natural Anionic Polymers

Many sulfated polysaccharides are available as natural biopolymers or their derivatives, some examples being heparin, dextran sulfate, mannan sulfate and chondroitin sulfate, but the applications are mainly medical (Voychek and Tan, 1993). One proposed for use in the water industry is the modified natural polymer lignin sulphonate, made by sulfonating low MW kraft pine lignin (Meister and Li, 1990).

Nonionic Polymers

Nonionic polymers include polyacrylamides, and naturally occurring polymers such as starches and cellulose derivatives.

Synthetic nonionic polymers

Nonionic polymers typically contain anionic and cationic groups, depending on the preparation conditions (Luttinger 1981,Vorchheimer 1981). ^{13}C nuclear magnetic resonance spectroscopy has been found to be the most reliable method for determining the degree of hydrolysis of concentrated polyacrylamide samples (Taylor and Nasr-El-Din 1994).

Natural Nonionic Polymers

Naturally occurring nonionic polymers include starches, galactomannans, cellulose derivatives, microbial polysaccharides, gelatins and glues. They are used as flocculants to aid solid-liquid separations, and vary in their structure, MW, biodegradability and ease of dissolution (Levine, 1981). Their main advantages are ready acceptance on health grounds and ease of biodegradation. Tannins have also received attention (Rice et al. 1964). The purification of drinking water with macerated seeds from the horseradish tree *Moringa oleifera* is of interest (Ndabigengesere et al. 1995).

CONCERNS RELATED TO POLYELECTROLYTE USAGE

The major concerns associated with the use of polyelectrolytes are related to potential health impacts from residual monomers (epichlorhydrin and acrylamide), and potential environmental impacts from residual polymers and monomers in water, wastewater, process residuals, or biosolids. These concerns have prompted several investigations into polymer toxicity and the development of analytical procedures for detecting residual polymer in the product water, and in wastes from water treatment processes.

Polymer toxicity

The types of anionic and nonionic polymers used in water treatment applications tend to be of low toxicity whereas cationic polyelectrolytes have been reported to be more toxic, especially to aquatic organisms. Concerns about potential toxicity have led Japan and Switzerland not to permit the use of polyelectrolytes in drinking water treatment, while Germany and France have set stringent limits. The health significance of possible contaminants has been reviewed in detail (Letterman and Pero 1990). The monomers are more toxic than the polymers (Criddle 1990). Limits on the level of monomer are strictly controlled, especially with acrylamide products, where, in most countries, the maximum allowable content of free acrylamide is 0.025%, and the residue in drinking water is limited to 25 μg/L. For PolyDADMAC the monomer content limit is 5% (Vitvitskaya et al. 1988). Recommended maximum doses of frequently used commercial polymers for drinking water production have been developed by the National Sanitation Foundation (NSF International 2001). The limits are outlined in Table 1.8. The PolyDADMAC value is based on a maximum carryover of polymer into the product water of 50 μg/L.

Table 1-8
Summary of Maximum Allowable Dosages of Organic Polyelectrolytes used in Drinking Water Applications

Polymer or monomer	Maximum allowable dose[1]	Maximum allowable residual in treated water
PolyDADMAC	50 mg/L	
EPI-DMA	20 mg/L	
Polyacrylamide	1 mg/L	
Acrylamide monomer	0.025% in polymer	25 μg/L

[1]Adapted from National Sanitation Foundation, 2004

In determining the toxicity of polymers to aquatic species there are different parameters compared to non-polymeric additives (Vitvitskaya et al. 1988). Synthetic polymers tend not to be readily absorbed by organisms and their toxicity may be substantially altered in the environment (Hamilton et al. 1994). Cationic polymers are rated at moderate to high toxicity, and are more toxic to aquatic organisms than anionic or nonionic polymers (Hall and Mirenda 1991). The cationic polyelectrolytes are detrimental to fish because of mechanical gill blockage that causes suffocation (Biesinger and Stokes 1986, Cary et al. 1987). This is greatly reduced by the addition of solids such as clays that are normally present in various forms in receiving waters and sediments, where they can strongly adsorb the cationic polyelectrolyte. Humic acids also affect the toxicity, reducing it by an order of magnitude at humic acid levels of 5 mg/L (Goodrich et al. 1991).

A study of the toxicity of cationic polyacrylamides of various molecular weights and CDs towards daphnia and minnows has shown that the toxicity is greatest for polymers of high CD, almost independent of chain length (Timofeeva et al. 1994). At lower CDs, the higher molecular weight polymers are more toxic in the case of daphnia, but the actual toxic level of 0.2 mg/L is well above the likely polymer concentration in a drinking water. The toxic level for minnows is 10 times greater. For anionic polyacrylamides the most toxic are those of longer chain length, but they are at least 100 times less toxic than the cationic versions. Cationic polymers disrupt yeast cells at a polymer dosages of about 70 mg/L, while anionic or non-ionic ones do not (Narita et al. 2001). Hydrophobic polylectrolytes tend to cause more cellular disruption than hydrophilic compounds.

A summary of the toxicity of polymers to freshwater organisms indicates that fish are more sensitive to cationic polymers, but algae are sensitive to anionic polymers because of the chelation of nutrient metal cations. This effect can be offset by the addition of Ca^{++} (WRc 1998). The presence of humic substances or clays reduces the bioavailabilty of organic polyelectrolytes and associated toxicity.

Residual polymer

In addition to polyelectrolyte residuals that may be present in coagulated and filtered water, polymer formulations used in water treatment processes can also re-enter the treatment system through recycled flow streams (backwash water and thickening basins) or become concentrated in process residuals (sludges). A variety of approaches have been applied to quantify residual polymers. Goppers and Straub (1976) used thin layer chromatography to detect residual polymer and monomer associated with a CPAM. In the analysis of a high molecular weight, low CD copolymer of acrylamide and sodium acrylate, separation by size exclusion chromatography and fragmentation by flash pyrolysis followed by gas chromatography/mass spectrometry (GC/MS) have been employed (Black et al. 1966). Size exclusion chromatography was also used in a study of residual non-ionic polyacrylamide in an alum sludge (Keenan et al. 1998). A number of impurities were present at very low levels in water treated with the commercial product, including the monomers acrylamide and sodium acrylate, and also hydroxypropionitrile and traces of isobutyronitrile from the initiator. The sensitivity was 10-20 μg/L.

A number of approaches for tracing residual polymer have been attempted as a means to confirm that all added polymer remains attached to particles that are removed prior to water use. One approach is to employ colloid titration against a polyelectrolyte of opposite charge, using dyes or fluorescent compounds as indicators (Wang and Shuster 1975; Parazak et al. 1987; Tanaka and Sakamoto 1993). The method is only effective for concentrations over 1 mg/L. A variation on this theme for cationic flocculants is to monitor turbidity by adding tannic acid to precipitate the polymer (Attia and Rubio 1975). A similar method measures the light absorbance at 680 nm following progressive additions of a standard solution of an o,o'-dihydroxyazo compound that participates in an association reaction with the polymer (Kirie et al. 1986).

In a review of methods available to determine cationic polyacrylamides, fluorescence spectroscopy was selected as the most promising analytical method (Howes and Harper 1998). Further method development is needed to achieve the desired detection limit of 50 μg/L, and it was felt that there should be more effort by polymer manufacturers to develop suitable methods. Size exclusion chromatography was also considered, but adsorption of polymer to the stationary phase resulted in analytical uncertainties. Another approach is based on a standard clay test (Burkert, 1970), where a calibration curve of polymer versus turbidity can be used to measure the content of residual polymer when its sample is tested on the suspension. However, the sensitivity is not high (Becker et al. 2000). A review of 17 groups of methods for determining polyacrylamides used in the petroleum industry has been published (Taylor and Nasr-El-Din 1994). The lowest detection limits were 10 and 20 μg/L, for SEC and fluorescence spectrometry respectively.

A number of workers have followed the course of flocculation with specially synthesized ^{14}C labeled polymer (Black et al. 1966). It has been used to show that dissolved organic matter reacts with the polymer before the turbidity particles (Narkis and Rebhun 1983), and that hydrolytic degradation of polyacrylamide is negligible under normal operating conditions (Hollander et al. 1981). Typically, polymers tend to partition onto solid particles (Anderson et al. 1993b). A fluorescently-labeled cationic polyacrylamide has been made by including choline in a Hofmann reaction on polyacrylamide (Tanaka and Odberg 1989).

More recently, PolyDADMACs have been made that contain 1-2% of an amine-functional monomer. These copolymers were reacted with a flurophore so that a fluorescent tag was formed within the polymer structure (Bennett et al. 2000). Residual PolyDADMAC could be detected fluorometrically at concentrations below 100 μg/L. Measurable amounts of residual polymer were always present in the treated water, especially at doses above or below the optimum.

Polymer degradation

Another concern about the use of organic polyelectrolytes is the potential for degradation during storage or during treatment. Degradation can occur during storage through a number of mechanisms. One of the major degradation pathways is hydrolysis. For example, hydrolysis of the ester links in cationic polyacrylamides is known to be pH and CD dependent (Lafuma and Durand 1989; Aksberg and Wagberg 1989; Smith-Palmer et al. 1994). Amide groups generally are much more stable.

The biodegradation of synthetic polymers is a slow process because most of the functional groups except amides are resistant (Satyanarayana and Chatterji 1993), as are the ester links in CPAMs (Soponkanaporn and Gehr 1989). Partial cleavage of the ester links under both aerobic and anaerobic conditions has been confirmed (Chang et al. 2001). The poly(acrylic acid) formed can be degraded in the natural environment, but only material of very low molecular weight (below 4000) seems to be affected (Lenz 1993). Biodegradability of the methacrylate chain can be facilitated by incorporating N-benzyl-4-vinylpyridinium units (Kawabata et al. 1994). A more biodegradable anionic polyelectrolyte in the form of sodium polyvinyloxyacetate has been prepared which cleaves to produce easily degradable poly(vinyl alcohol) (Matsumura et al. 1998).

Natural polymers based on polysaccharides are readily degraded at the ether sites and and proteins are degraded at the amide sites (Satyanarayana and Chatterji 1993). Synthetic analogues of natural polymers that are easily degraded, such as poly(aspartic acid) and acrylic acid grafts on polysaccharides, are seen to have a future as biodegradable scale control agents (Low and Koskan 1993; Dezinger et al 1990).

Storage practices for polyelectrolytes are variable, depending on the form of the polyelectrolyte (liquid or powder), the rate of use, and plant operating practices (MacPhee and Cornwall, 2002). In some cases, polyelectrolytes are stored under controlled temperature environments whereas in other situations, they are stored under ambient environmental conditions. Limited information is available about the extent of on-site degradation that occurs during storage.

Formation of Disinfection By-Products

Perhaps one of the main concerns about the use of organic polyelectrolytes is the potential for formation of disinfection byproducts. With the emphasis of water treatment applications on control of residual levels of organics, the addition of organics to water may, in fact, be counterproductive. A summary of information on the potential for organic polyelectrolytes to act as disinfection byproduct precursors is presented here.

Reactions with chlorine

The major concern about reactions with chlorine is the potential for formation of disinfection byproducts (DBPs). Several studies have been conducted with an emphasis on the formation of trihalomethanes.

Cationic polyelectrolytes. In general, it has been reported DBP formation from oxidation reactions with cationic polymers such as PolyDADMAC and EPI-DMA, are limited in contrast to the amount of DBP formation associated with NOM reactions (Fielding et al. 1999, Alekseeva and Khromchenko 1988, Kaiser and Lawrence 1977, Chang et al. 1999, Fielding et al. 1999). A summary of reported data on trihalomethane formation from reactions of cationic polyelectrolytes with chlorine is shown in Table 1.9. The highest concentrations of trihalomethanes resulted from chlorine reactions with a high molecular weight cationic polyacrylamide (Soponkanaporn and Gehr, 1989). They reported that the rate and extent of polymer reaction increased with increases in pH and temperature. At pH 7 with 10 mg/L of polymer, 28 μg/L of chloroform was produced after 20 days (2.8 μg/mg); however, at pH 9 the chloroform produced was 58 μg/L (5.8 μg/mg).

The major variables that appear to influence the formation of disinfection byproducts are the relative dosages of chlorine and polymer and contact time. A comparison of trihalomethane formation from 72 hour exposure of EPI-DMA to high dosages of chlorine at neutral pH is shown in Figure 1.3 (data from Fielding et al. 1999). Normalized data are shown in Figure 1.4 (correlation coefficient, r^2 is 0.85). As shown, about 0.6 μg of trihalomethanes are formed per mg of EPI-DMA. In comparison to THM formation from NOM, this amount is negligible at the dosage ranges used in typical water treatment applications. Fielding et al. (1999) also found that significant quantities of AOX (192 μg/L) form from reactions of chlorine and EPI-DMA, suggesting that additional chlorinated compounds may be formed from EPI-DMA (Fielding et al. 1999).

Limited data are available on the formation of disinfection byproducts other than trihalomethanes. Fielding et al. (1999) reported that when EPI-DMA reacts with chlorine, the DPBs formed included chloroform (3.7 μg/L), dichloroacetic acid (19.9 μg/L), trichloroacetic acid (12.8 μg/L), formaldehyde (61 μg/L), glyoxal (14.6 μg/L) and methylglyoxal (23.9 μg/L). Possible precursors to disinfection byproducts include unreacted monomers and organic impurities in the polymer such as epichlorohydrin and dimethylamine, glycidol, ethylenediamine and various chlorinated aliphatic alcohols present as contaminants in the epichlorohydrin (Letterman and Pero, 1990).

Table 1-9.
Formation of trihalomethanes from cationic polyelectrolyte reactions with chlorine at neutral pH

Polyelectrolyte Type	Polyelectrolyte Dose, mg/L < 2	Polyelectrolyte Dose, mg/L 5	Polyelectrolyte Dose, mg/L 10	Chlorine Dose, mg/L < 5	Chlorine Dose, mg/L 5-9	Chlorine Dose, mg/L ≥ 10	Exposure time, hours ≤ 4	Exposure time, hours 4-24	Exposure time, hours ≥ 24	Other Factors	TTHM, µg/L	Reference
PolyDADMAC												
			x	x			x			Dark	1.2	Kaiser and Lawrence 1977
			x	x					x	Dark	1.5	Kaiser and Lawrence 1977
			x	x			x			UV	3.1	Kaiser and Lawrence 1977
	x				x		x			NOM	20	Alekseeva and Khromchenko 1988
		x			x		x			NOM	36	Alekseeva and Khromchenko 1988
			x	x					x		2	Chang et al. 1999
			x	x					x		5	Chang et al. 1999
	x					x	x				1.0	Fielding et al. 1999
Epichlorhydrin Dimethylamine (EPI-DMA)												
			x		x				x		3.6	Feige et al.1980
			x			x			x		6.3	Feige et al.1980
			x		x				x		8.5	Feige et al.1980
			x			x			x		9.9	Feige et al.1980
		x				x		x			3.7	Fielding et al. 1999
Cationic Polyacrylamide (CPAM)												
			x			x			x	pH 7	28	Soponkanaporn and Gehr 1989
			x			x			x	pH 9	58	Soponkanaporn and Gehr 1989

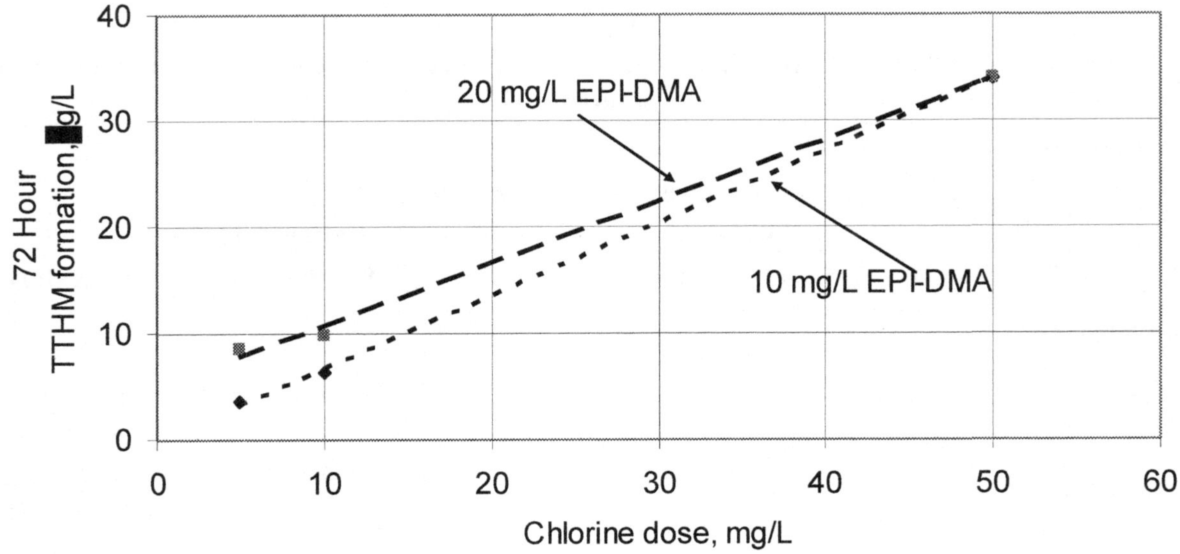

Figure 1-3 Comparison of 72 hour trihalomethane formation from reaction of chlorine with 10 or 20 mg/L EPI-DMA (Data from Fielding et al. 1999)

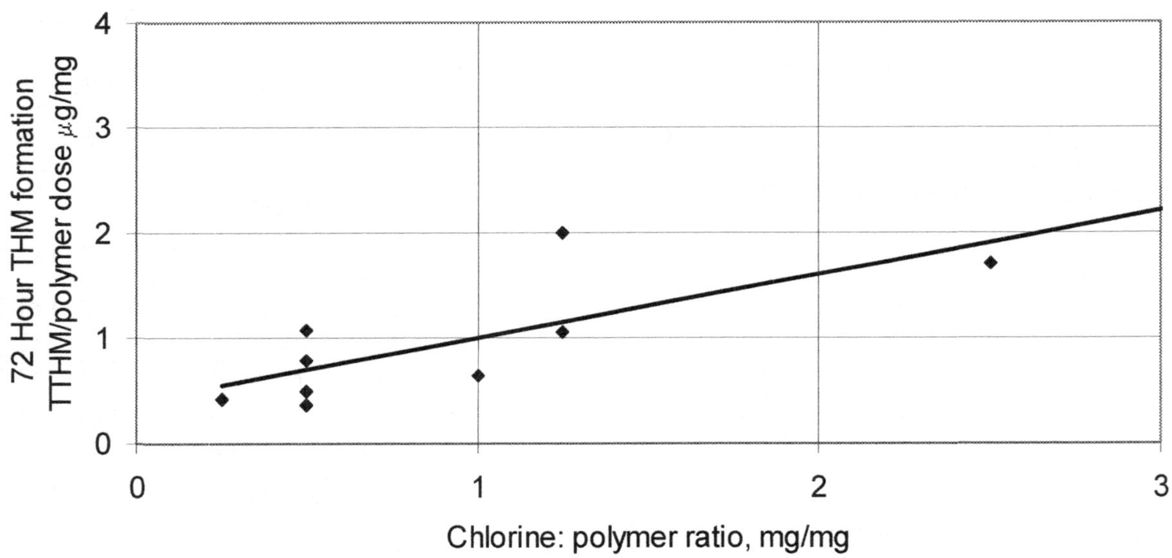

Figure 1-4. Influence of the ratio of chlorine to polymer dose on the 72 hour trihalomethane formation from EPI-DMA (Data from Fielding et al. 1999).

Polydimethylaminoethyl methacrylate (PDMAEMA) has been tested for chloroform production in laboratory studies on neutral colored waters treated with alum and then 6 mg/L of chlorine for two hours. A negligible increase in chloroform was observed (from 76 to 80 μg/L) when 0.5 mg/L of PDMAEMA was added (Alekseeva and Khromchenko, 1988). However, when chlorine was added after polymer addition, chloroform formation decreased from 109 to 24 μg/L for one water and from 123 to 10 μg/L for another. It is possible that the addition of a cationic polymer allows for more efficient removal of the natural DBP precursors, thereby reducing the overall DBP formation potential. While amino compounds are vulnerable to chlorine attack, the tertiary amino centers of the type in PDMAEMA are less reactive than the primary (or secondary) amines.

For two colored waters of neutral pH treated with alum and then 6 mg/L of chlorine for two hours, there was no significant difference in the production of chloroform on addition of 0.5 mg/L of Polyethyleneimine (PEI) (Alekseeva and Khromchenko, 1988). When polymer was added before chlorination, the chloroform produced decreased to 33 μg/L compared to 109 μg/L in the absence of PEI for one of the waters; for the other water the data were 32 and123 μg/L. Such behavior is to be expected since removal of the colored compounds is more efficient with the cationic polymer present. On adding greater amounts of PEI to another water, there was a decrease from 110 μg/L with no polymer present to a minimum production of 17 μg/L at a 1 mg/L dose of polymer, gradually rising to 65 μg/L at a 5 mg/L dose (13 μg/mg).

Anionic polyelectrolytes. The formation of DPBs from chlorine reactions with APAMs has been reported by several researchers. A summary of trihalomethane levels detected after exposure of APAMs to chlorine is shown in Table 1.10.

Table 1-10.
Formation of trihalomethanes from anionic polyacrylamide reactions with chlorine at neutral pH

APAM Dose, mg/L			Chlorine Dose, mg/L				Exposure time, hours			TTHM, μg/L	Reference
≤ 5	10	40	≤ 5	5-20	20	50	≤ 4	4-24	72		
		x			x				x	70	Feige et al. 1980
		x				x			x	94	Feige et al. 1980
x			x						x	24	Feige et al. 1980
x						x			x	58	Feige et al. 1980
	X				x		x			1.2	Fielding et al. 1999
x				x				x		7 to 8	Mallevialle et al. 1984

Fiege et al. (1980) exposed a 40 mg/L solution of a commercial APAM to chlorine dosages of 20 and 50 mg/L neutral pH and 20° C for three days (Feige et al., 1980). Chloroform production was 70 (1.75 μg/mg at a chlorine:polymer ratio of 0.5 mg/mg)) and 94 μg/L (2.35 μg/mg at a chlorine:polymer ratio of 1.25) at the two dose levels. Trace amounts of bromodichloromethane (0.3 to 0.9 μg/L), were attributed to the presence of brominated compounds in the original polymer, since the reaction was carried out in purified water and three other polymers did not yield any of this by-product. When a polymer dose corresponding to the maximum allowed by the US EPA was used (4 mg/L), plus chlorine doses ranging from 5 to 50 mg/L, chloroform concentrations ranged from 24 μg/L (6 μg/mg at a chlorine:polymer ratio of 1.25 mg/mg) to 58 μg/L (14.5 μg/mg at a chlorine:polymer ratio of 12.5 mg/mg). Both sets of tests gave by-product levels up to twice those found for EPI/DMA, even at much lower polymer concentrations. It is possible that other impurities present in the commercial product may have contributed to DBP formation.

For a polymer of about 10,000,000 amu molecular weight and a CD of 10-12%, by-product formation was attributed mostly to the low molecular weight material present (Fiessinger et al. 1983). The THM levels obtained from the same APAM in another report were given as 7-8 μg/L when 1 and 5 mg/L samples of the acrylamide/sodium acrylate copolymer were both reacted with 2 and 10 mg/L of chlorine for one to 24 hours (Mallevialle et al., 1984). Although the polymer produced little THM, the contaminants were capable of forming much larger amounts, particularly acrylamide monomer, which yielded 800 μg/mg. This shows the importance of adequate quality control during manufacture. An acrylamide specification limit of 500 mg/kg of polymer would yield 0.4-2.0 μg/L of THM at the polymer doses cited. Other contaminants such as acrylic acid and hydroxypropionitrile, an impurity in the acrylamide, were present at much lower specified limits of 50 and 100 mg/kg of polymer respectively. They also form THMs, but in smaller quantities.

Chloroform production from colored waters (COD 25-29 mg/L) treated with alum and then 6 mg/L of chlorine for two hours was relatively unchanged at 74 versus 76 μg/L when 0.5 mg/L of APAM was later added and chloroform amounts were determined immediately after filtration. Two hours later the respective levels were 92 and 97 μg/L (Alekseeva and Khromchenko 1988). When the polymer was added before chlorination the chloroform formation decreased from 109 to 76 μg/L for one water and from 123 to 79 μg/L for another, indicating that build up of alum flocs enhances the removal of the colored compounds and suggesting that the NOM is the main source of chloroform. Higher dosages of polymer yielded reduced chloroform production from 140 μg/L with no polymer present to a minimum production of 20 μg/L at a 3 to 5 mg/L polymer dose. This is confirmation that the APAM polymer itself is not a precursor for chloroform formation under these conditions.

Chain scission occurred when an APAM of about 16,000,000 amu and CD of 10% was reacted with 70 mg/L of chlorine. After a three hour reaction time, the molecular weight decreased to 30,000 with further reduction in size associated with longer reaction times (Aizawa et al., 1991). Reductions in molecular size can interfere with the effectiveness of the APAM as a flocculant.

The most recent work reported minimal chloroform formation (1.2 μg/L) from APAM, using 10 mg/L of solid polymer and 25 mg/L of chlorine for 4 hours at pH 7 (Fielding et al. 1999). The AOX was 21 μg/L and other by-products included dichloroacetic acid (9 μg/L), trichloroacetic acid (13 μg/L), formaldehyde (34 μg/L) and glyoxal (1.9 μg/L). Similar results were obtained for an emulsion form of the polymer, although the AOX was much higher at 69 μg/L. The contaminants in APAMs other than monomers include hydroxypropionitrile and isobutyronitrile (Letterman and Pero 1990). Chlorination of monomers and impurities did not yield readily detectable amounts of chlorination by-products (Fielding et al. 1999).

In contrast to APAMs, anionic polystyrene is much more reactive with chlorine. An increase of about 20% in chloroform production was observed when colored waters were treated with alum, 0.5 mg/L of polystyrenesulfonate (PSS), and exposed to chlorine for two hours (Alekseeva and Khromchenko 1988). When the polymer was added before chlorination, chloroform formation increased from 109 to 189 μg/L for one water and from 123 to 203 μg/L for another. Polystyrene or its contaminants appear to be substantial contributors to chloroform production, as might be expected from the chlorine reactivity of aromatic compounds. However these polymers are not used in potable water treatment.

Other polyacrylamides of the anionic and non-ionic type present less of a problem, with THM formation at trivial levels compared to the amounts produced by NOM (Alekseeva and Khromchenko 1988, Feige et al. 1980, Mallevialle et al. 1984, Fielding et al. 1999). These polymers have contributions equivalent to the amount produced by the monomer present.

Reactions of ozone with organic polyelectrolytes

Ozone is a powerful oxidant and it can oxidize organic polyelectrolytes. The typical reaction products from ozonation of polymer when used at levels typical of water treatment include formaldehyde. At polymer levels of 10 mg/L, ozonation of PolyDADMAC produced 36 μg/L of formaldehyde whereas EPI-DMA produced 62 μg/L (Fielding et al. 1999). Formic acid and other aldehydes were also produced. For APAMs, ozone can completely destroy polymers if adequate contact time is available. However, under typical conditions, formaldehyde is the major ozonation byproduct. Mutagenicity tests on an ozonated APAM have proved negative.

Reactions of polyelectryolytes due to chlorine addition after ozone.

Chlorination after ozonation can be problematic when there has been no removal of low molecular weight compounds that are produced from ozonation. Chlorinated byproducts can form from chlorine reactions with the low molecular weight compounds. For polyDADMAC, an AOX of 435 μg/L was obtained following excessive doses of both reagents (Stockham and Morran, 2000). A major product was chloropicrin (197 μg/L); chloroform, dichloroacetic acid and formaldehyde were each produced at levels above 50 μg/L, as well as smaller amounts of many other chlorinated species and ketoacids.

Polyelectrolyte reactions with permanganate.

Permanganate, like ozone, can oxidize organic polyelectrolytes depending on the dose and contact time. Reactions of permanganate with PolyDADMAC resulted in a chain shortening and loss of charge density (Levine et al., 1998). Similar to ozonation, pretreatment with permanganate prior to chlorine may contribute to the formation of disinfection by-products.

Polyelectrolyte reactions with ultraviolet light.

Caulfield et al. (2003) described the effect of light irradiation and thermal treatment on PAM. The polymer was stable under fluorescent light and did not release any detectable acrylamide monomer in hot aqueous solutions after 15 days at 95°, although hydrolysis of amide groups was observed. After 10 days about 10% of the amide groups had been converted to carboxylic acid sites. Under UV irradiation small levels of monomer were detected after 10 days, but they were generally below 50 parts per million repeat units in the polymer. There was a decrease in the molecular weight, indicating that the monomer release was because of random chain scission, not an unzipping of the polymer chain.

Kaiser and Lawrence (1977) found that exposing concentrated EPI-DMA polymer at 100 mg/L with 10 mg/L of chlorine to UV for one hour resulted in production of 13.8 μg/L of chloroform (0.14 μg/mg).

A cationic polyacrylamide (CPAM) was irradiated with UV light at a pH of 3 for 1 hour and the trihalomethane formation potential increased six fold (Soponkanaporn and Gehr 1989), possibly due to the presence of acrylate or acrylamide monomers. Acrylamide is known to be a potent trihalomethane precursor yielding 800 μg of chloroform per mg of acrylamide (Mallevaille et al., 1984). Other potential contaminants in the polymer include the monomers, hydroxypropionitrile, and isobutyronitrile (Letterman and Pero 1990).

Potential Formation of N-nitrosodimethylamine (NDMA)

N-nitrosodimethylamine (NDMA), a known animal carcinogen of 1000 times the potency of THMs, has been detected in several potable water supplies though the raw water had no detectable NDMA, indicating NDMA formation during treatment (Davis et al, 2000; Andrews and Taguchi 2000, Valentine et al. 2004). Polymers and alum have been identified as likely sources of NDMA precursors (Davis et al. 2000, Child et al. 1991, Kruger et al. 2000). The monomers used to form PolyDADMAC are dimethylamine and allyl chloride. Other potential sources of NDMA include reactions of the seed fungicide Thiram nitrite and hydrogen ions (Graham et al. 1996). In chlorinated wastewater effluents, NDMA formation has been reported to exceed 100 ng/L (Njam and Trussell 2001). A suggested reaction scheme is via a nucleophilic attack of dimethylamine (or any secondary amine) on monochloramine, to form a hydrazine intermediate that is rapidly oxidized to the corresponding N-nitrosamine (Mitch and Sedlak 2001).

NDMA formation has been documented from the reaction of polyDADMAC with chloramine in water treatment (Wilczak et al., 2003). The polymer was the only significant source of NDMA precursors and the amount produced increased with polymer dose and longer chloramines contact times. Pre-ammoniation or simultaneous addition of chlorine and ammonia resulted in higher levels of NDMA. Negligible amounts were formed using free chlorine at doses typical for disinfection. Decreasing the polyDADMAC dose from 0.8-1.2 to 0.3 mg/L and maintaining free chlorine contact before chloramine formation helped reduce NDMA levels from over 10 to around 2 ng/L. The yield of NDMA also increased linearly for EPI-DMA polymers in chlorinated raw surface waters (Kohut and Andrews 2003). The doses of polymer and chlorine at several mg/L were relatively high and significant amounts of nitrogen compounds were also present.

ECONOMICS OF USING POLYELECTROLYTES

The use of polyelectrolytes has advantages in terms of improved turbidity removal and reduced sludge production, however, they tend to be relatively costly chemicals compared to inorganic coagulants. Few studies have been published that focus on the economics of using polyelectrolytes. An evaluation of the performance and economics of 23 different organic polyelectrolytes in treating a turbid Texas water source was conducted to identify a polymer to reduce filter loading by improving turbidity removal (Tarquin et al. 1998). To be economically viable, the savings in filter backwash costs should off-set the costs of polymer addition. A 5.4 hour increase in filter run time equated to an 18% reduction in the volume of backwash water required, yielding a savings of $4.32 per million gallons of product water. Depending on the cost of polymer and its performance at a dosage of 0.5 mg/L, the net savings varied from $2.56 to -$3.20 per million gallons.

In South African, changing from alum to a polymeric coagulant resulted in a 30% decrease in the unit cost over a three-year period despite an annual inflation rate of about 10% (Nozaic et al., 2001). There was also a small reduction in power cost as the polymer dose was 10-20% of that for alum, allowing the use of smaller dosing pumps that required less power. As a further bonus the organic polymer was less aggressive than inorganic coagulant, resulting in reduced maintenance costs. The volume of sludge produced was also reduced by conversion to the polymeric coagulant.

SUMMARY

Polyelectrolytes are widely used in water treatment applications. Studies have been conducted to evaluate potential formation of disinfection byproducts. Based on the literature review reported here it is important to evaluate the potential for advanced oxidation reactions to occur in the presence of ultraviolet light and oxidant chemicals. A summary of the results of the polyelectrolyte-oxidant interactions identified through the literature survey is given in Table 1.11.

Table 1-11
Summary of polyelectrolyte-oxidant interactions identified in the literature review.

Polyelectrolyte Classification	Chemical composition	Oxidant reactions
Cationic	Polydiallyldimethylammonium chloride	Chlorine, Ozone, Permanganate
	Poly-2-hydroxypropyl-1-N,N-dimethylammonium chloride	Chlorine, Ozone
	Cationic polyacrylamides	Chlorine
	Polydimethylaminoethyl methacrylate	Chlorine
	Polyethyleneimine	Chlorine
Anionic	Anionic polyacrylamides	Chlorine, Ozone
	Anionic polystyrene	Chlorine
Non-Ionic	Polyacrylamide	Ozone/UV

CHAPTER 2: METHODS AND MATERIALS

The focus of this project is on chemical interactions that occur during drinking water treatment with a specific emphasis on polyelectrolyte interactions. The specific project objectives are:

1. Evaluate current patterns of chemical use for drinking water production with respect to the types of chemicals used, chemical dosages, chemical addition practices, potential chemical interactions, and operational measures of chemical effectiveness.
2. Conduct bench-scale testing to evaluate polyelectrolyte interactions and the impacts of these interactions on process performance, water quality, and operations.
3. Assess the degree to which exposure to ultraviolet light impacts polyelectrolyte reactions with oxidants
4. Conduct a laboratory study to determine the effect of precoating filter media with various polyelectrolytes on filtration efficiency and filter backwashing effectiveness.

The methods and materials used for this project are provided in this chapter in terms of the individual project objectives.

SURVEY DEVELOPMENT

The goal of this component of the project was to document chemical use patterns in drinking water production. A survey was developed to compile information on the types of chemicals, dosages, point of addition, mixing, methods of monitoring residuals, and sludge production. The primary sources of information used in compiling information included AWWA databases; State water quality agencies, and water utilities.

The survey was designed to identify facilities where polyelectrolytes are in use, evaluate the types of polyelectrolytes in use, identify combinations of chemicals that are used before or after polyelectrolyte addition, determine the potential for sunlight interaction with polyelectrolytes.

An INTERNET-based survey was developed that is linked to an ACCESS database. The main components of the survey are listed in Table 2.1. Several approaches were used to encourage participation in the survey. E-mail contact information was obtained from utility websites. Each utility that was contacted by e-mail was assigned a user name and password and invited to complete the survey on-line. The survey responses are directly entered into a database that is resident on the server.

Table 2-1
Summary of major elements of INTERNET survey

Component	Information obtained
General	Name of utility, location, size, age, private or public, water source(s)
Source water quality	Characteristics of water source(s) including pH, alkalinity, turbidity, TOC, conductivity.
Treatment information	Operating characteristics of treatment units including size, loading rates, sequence, exposure to sunlight
Chemical information	Details of the chemicals added during treatment including vendor, storage conditions, point(s) of addition, dosages, monitoring, pH at addition point, seasonal variations, etc.

The survey pages are written as Active Server Pages (ASP). ASP allows web sites to be dynamic and interactive. Each page contains two parts: programmatic code and embedded HTML. The programmatic code is written in a *scripting language* called *Vbscript*. The survey is hosted on a web server at the University of South Florida and the participants enter information using their web browsers. The website for the survey is: http://ceweb.eng.usf.edu/login.asp/.

Each user is assigned a user name and password to log on to the survey and prevent unauthorized use. Upon the first visit to the survey website, users are asked to complete a general information form. This form collects Utility and Contact information. Required fields are labeled in red. After the form is completed, the user submits the information and is presented with a summary of the submitted information to check its validity. Continuation links are provided on each page.

After completing the General Information form, users are routed to the Main Menu where the survey branches into three sections: water sources, treatment processes, and chemical use information. Each section is self-contained and subdivided into individual pages to facilitate data entry. The first page of each section asks for general information and the answers to the general information questions are linked to specific follow-up questions. An excerpt from the introductory page on source water is shown in Figure 2.1. Users select the types and quantities of water sources that serve the facility and then are asked specific questions pertaining to each source.

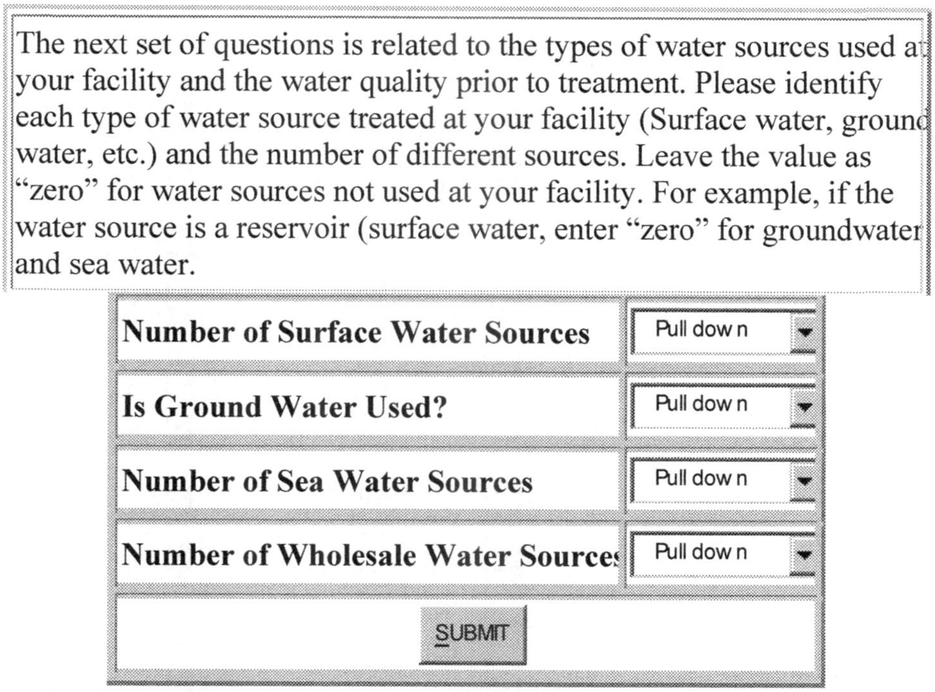

Figure 2-1. Excerpt from introductory survey page for on water sources from On-Line Water Treatment Survey.

After entering the source information, users are queried for treatment information. The first page of the treatment section asks the user to select the types of treatment that are used from an alphabetical list. The treatment selection list is shown in Figure 2.2. Depending on the treatments that are selected, follow-up questions are asked. Follow-up questions on Coagulation/Flocculation are shown in Figure 2.3. Similar types of questions are asked for each treatment unit..

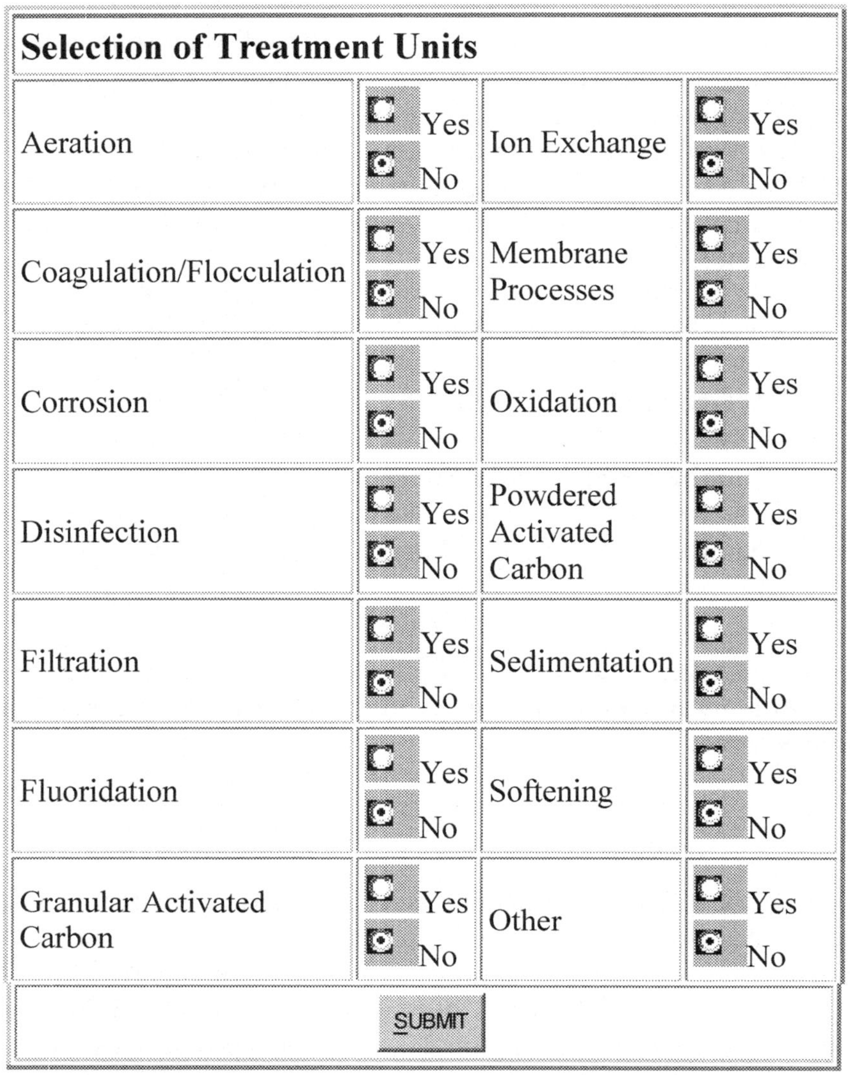

Selection of Treatment Units			
Aeration	Yes / No	Ion Exchange	Yes / No
Coagulation/Flocculation	Yes / No	Membrane Processes	Yes / No
Corrosion	Yes / No	Oxidation	Yes / No
Disinfection	Yes / No	Powdered Activated Carbon	Yes / No
Filtration	Yes / No	Sedimentation	Yes / No
Fluoridation	Yes / No	Softening	Yes / No
Granular Activated Carbon	Yes / No	Other	Yes / No
	SUBMIT		

Figure 2-2. Alphabetical list of treatment units excerpted from On-Line Water Treatment Survey.

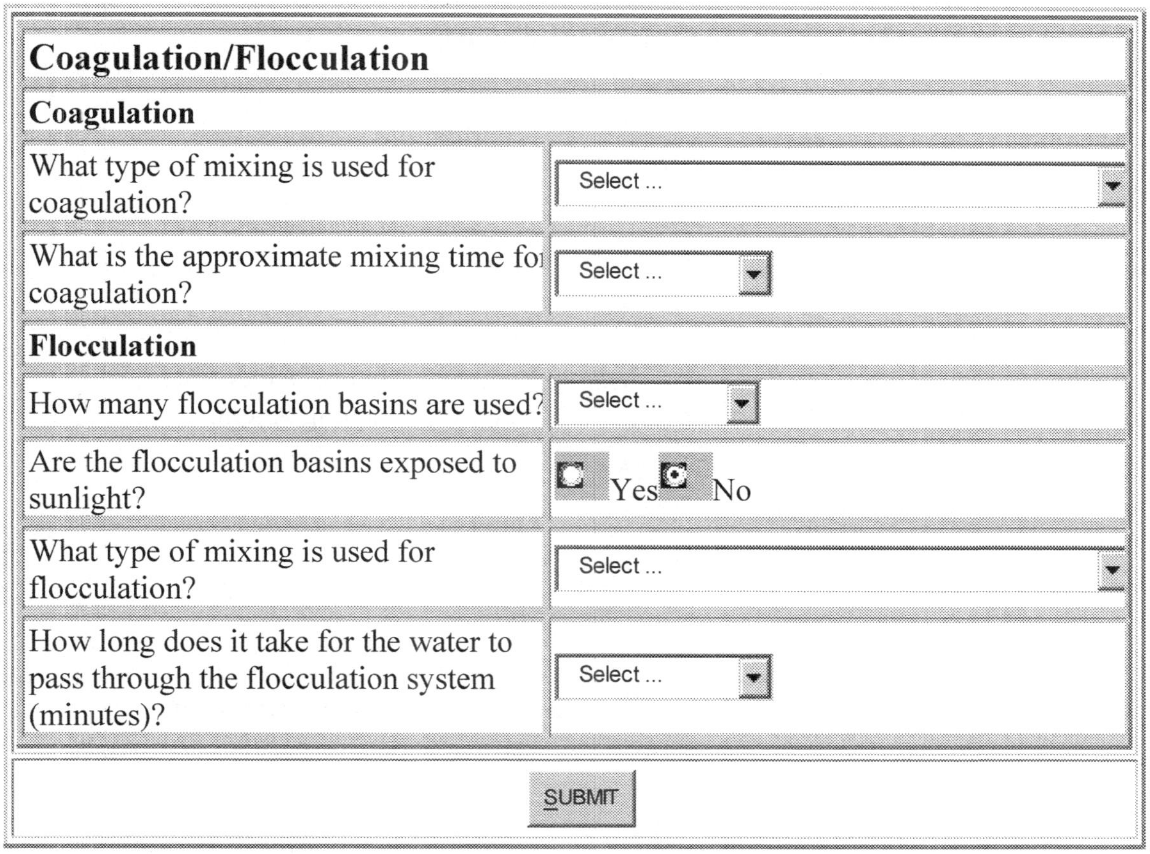

Figure 2-3. Questions relating to coagulation/flocculation excerpted from On-Line Water Treatment Survey.

The most detailed part of the survey is the chemical use information. Users select the chemical groups in use as shown in Figure 2.4. Once the categories of chemicals have been selected, specific chemicals are selected. Examples of the chemical choices for Coagulants/Flocculants/Filtration are shown in Figure 2.5; chemical choices for Disinfection/Oxidation/Taste and Odor Control are shown in Figure 2.6; and chemical choices for pH control/Alkalinity adjustment are shown in Figure 2.7.

Chemical Groups			
Algae Control	Yes No	Fluoridation	Yes No
Coagulants/Flocculants/Filtration	Yes No	pH Control/Alkalinity Adjustment	Yes No
Corrosion Control/Antiscalants/Sequestration	Yes No	Reducing Agents/Dechlorinating Agents/Quenching Agents	Yes No
Disinfection/Oxidation/Taste and Odor Control	Yes No	Other	Yes No
SUBMIT			

Figure 2-4. Excerpt from Introductory Page for Chemical Use Information.

Identify the chemicals used for Coagulation/Flocculation/Filtration.	
Alum	Yes No
FerricChloride	Yes No
FerricSulfate	Yes No
FerrousSulfate	Yes No
Polyelectrolyte	Yes No
PACl	Yes No
PASS	Yes No
PFS	Yes No
Other	Yes No

Figure 2-5. Chemical selection for Coagulation/Flocculation/Filtration chemicals excerpted from On-Line Water Treatment Survey.

Identify the chemicals used for Disinfection/Oxidation/Taste and Odor Control.	
Chlorine	Yes No
Chloramine	Yes No
ChlorineDioxide	Yes No
Ozone	Yes No
PotassiumPermanganate	Yes No
Other	Yes No

Figure 2-6. Chemical selection for Disinfection/Oxidation/Taste and Odor Control chemicals excerpted from On-Line Water Treatment Survey.

Identify the chemicals used for pH Control/Alkalinity Adjustment.	
CalciumCarbonate	Yes No
CalciumHydroxide	Yes No
CalciumOxide	Yes No
CarbonDioxide	Yes No
HydroChloricAcid	Yes No
Lime	Yes No
PotassiumHydroxide	Yes No
SodiumCarbonate	Yes No
SodiumHydroxide	Yes No
SulfuricAcid	Yes No
Other	Yes No

Figure 2-7. Chemical selection for pH Control/Alkalinity Adjustment excerpted from On-Line Water Treatment Survey.

Depending on the chemicals selected, the survey then queries the user for detailed information about the use of each chemical including questions pertaining to dosage, storage, make-up water, and monitoring. The questions are specific for each chemical. For example, if the user selects polyelectrolytes, then they are asked to identify the treatment units in which polyelectrolytes are used as shown in Figure 2.8.

Polyelectrolyte Treatment Units	
Coagulation	Yes No
Flocculation	Yes No
Filtration	Yes No
BallastedSedimentation	Yes No
DissolvedAirFlotation	Yes No
FilterBackWashTreatment	Yes No
Other	Yes No
CONTINUE	

Figure 2-8. List of potential uses for polylectrolytes excerpted from On-line Water Treatment Survey

Based on the treatment units that are selected, the user is queried for specific information. The specific questions that relate to polyelectrolyte use for coagulation are shown in Figure 2.9. In cases where all of the requested information is not available, users can skip questions.

All data from the survey are automatically compiled into an ACCESS database that can be used to analyze data trends and conduct statistical analyses. The database houses information on four types of information: utilities, water sources, treatment types and chemicals. Since information is associated with each utility, all of the records in all of the tables are referenced by a UtilityNum. This is a unique identifier for each utility that allows all of the records in the database to be traced back to a specific utility. Within the database, there are four tables on water source and each water source type is prefixed with the letters src. There are 15 tables on treatment types and each treatment table is prefixed with the letters tr. The bulk of the survey is devoted to chemical use information and there are 52 chemical use tables in the database with a prefix of chm. All other tables are used as look-up tables or for use by the web site application.

.Manipulating the data in the survey is accomplished by generating queries within the ACCESS database. Queries are used to extract data with similar attributes. For example, all of the data on chemical use in filters can be extracted using a query and then sorted by the type of filter, geographical location, type of source water, or any other category that is appropriate.

Polyelectrolyte	Coagulation
What quality control tests are performed before the Polyelectrolyte is used? ☐ None ☐ Monomer analysis ☐ UV scan ☐ Viscosity ☐ Metals analysis ☐ TOC analysis ☐ Other	
What is the tradename of the polymer?	Name
What is the vendor/manufacturer name?	Name
What form of the polyelectrolyte is applied?	○ Solid ○ Liquid
How is the polyelectrolyte stored?	Select one...
What is the average length of time that the polyelectrolyte is stored on site?	Select one...
If the Polyelectrolyte is liquid, what is the percent of active polyelectrolyte as received (%)?	Select one...
What is the chemical composition of the Polyelectrolyte?	Select one...
What is the dominant charge of the polyelectrolyte?	○ anionic ○ cationic ○ nonionic
What is the nominal molecular weight of the polyelectrolyte (from MSDS sheet)?	Select one...
What is the charge density of the polyelectrolyte (from Manufacturer)(meq/g)?	Select one...
Where in the treatment process is the polyelectrolyte added?	Select one...
What type of feed equipment is used?	
What type of mixing is used?	Select one...
What type of makeup water is added?	Select one...
How is the polyelectrolyte dose selected?	Select one...
What is the average dose of polyelectrolyte (mg/L)?	Select one...
How is the polyelectrolyte monitored?	Select one...
If Polyelectrolyte is not used all year, select the month(s) that they are used. ***Hold down the Ctrl key to select multiple months***	Select one... All Year January February
SUBMIT	

Figure 2-9. Questions that relate to polyelectrolyte use in coagulation excerpted from On-line Water Treatment Survey.

ANALYTICAL METHOD DEVELOPMENT

The focus of this project is to identify the potential for interactions between polyelectrolytes and other chemicals in the context of drinking water treatment. The goal is to quantify the interactions in terms of potential effects on the polymer and potential impacts on overall process performance. The test methods were designed to evaluate:

a. the fate of polyelectrolytes that are added to water that contains residual concentrations of other chemicals and
b. the extent to which residual (unreacted) polyelectrolytes might react with chemicals that are added downstream of polyelectrolyte addition.

All testing was developed to test chemicals under chemical addition and mixing regimes that simulate conditions in water treatment facilities. Initially, a simple water matrix was used to conduct a series of screening tests on each chemical combination. Chemicals that yielded positive reactions in the screening tests were subjected to follow-up testing in a surface water source. In each case, bench-scale testing was conducted using a jar-testing approach.

For this project, several tests were conducted to identifying changes in polyelectrolyte characteristics. An overview of the tests performed is given in Table 2.2

Table 2-2
Overview of methods for evaluation of polyelectrolyte properties.

Method	Rationale
Capillary electrophoresis	Separation of polyelectrolytes based on molecular size and electrophoretic mobility
Size exclusion chromatography	Separation of polyelectrolytes based on molecular size
Ultrafiltration	Physical separation of polyelectrolyte based on molecular weight cutoff of membranes
Charge density	Evaluate the influence of various chemical combinations on the charge density of polyelectrolytes
Disinfection byproduct formation potential	Evaluate the formation of DBPs from oxidant-polyelectrolyte reactions
Time to filter test	Develop a relative measure of the performance of polyelectrolytes based on the particle size and filterability of flocculated suspensions.

A set of polymers was selected to use for testing each analytical approach. A list of the polymers used for method development is given in Table 2.3. The types of screening tests that were used included measurement of the absorption spectra for different concentrations of each polymer using an ultraviolet/ visible light spectrophotometer. Other tests that were run included charge density, streaming current, permanganate reactivity, and UV/persulfate oxidation.

Table 2-3
List of polymers used for analytical method development

Polymer name	Type of polymer	Chemical composition	Specific gravity	Types of analyses
CATFLOC TL	cationic	PolyDADMAC	1.02 to 1.04	UV/VIS spectra, charge density, permanganate reactivity, TOC
POL-E-Z 675	anionic	"Multicomponent liquid"	1.04	UV/VIS spectra, charge density, permanganate reactivity, TOC
HYPERFLOC AF 300	anionic	Polyacrylamide	0.8	UV/VIS spectra, charge density, permanganate reactivity, TOC
SUPERFLOC N-1986	nonionic	Polyacrylamide	1.0	UV/VIS spectra, charge density, permanganate reactivity, TOC
SUPERFLOC C-572	cationic	EPI-DMA	1.14 to 1.18	UV/VIS spectra, charge density, permanganate reactivity, TOC

Indirect Tests

Several indirect tests of residual polyelectrolyte were conducted to identify the most appropriate measurements for each set of chemical interaction tests. The indirect tests included colloid titration against a polyelectrolyte of opposite charge, measurement of light absorbance at 680 nm following progressive additions of a standard solution of an 0,0'-dihydroxyazo compounds that participates in an association with the polymer (Bolto 1995, Hoog 1984, Gregory and Lee 1990). Anionic dye complexation was effective for measurement of cationic polyelectrolytes in water. Changes in the light absorbance of a standard dye solution (Ponceau S dye) are proportional to the mass of polyelectrolyte complexed with the dye (Hanasaki et al. 1985). This method was effective for concentrations greater than about 0.5 mg/L. Charge density measurements were also conducted (Dentel et al. 1993, Tiravanti et al. 1985, Van Leeuwen et al. 1997).

Ponceau Dye Test Methods

The Ponceau Dye Test method involves mixing a 50 mL sample of water containing polyelectrolyte with sulfuric acid (0.5M), Ponceau S dye (200 mg/L), and dichloromethane. Samples are mixed for 2 minutes and allowed to stand for 15 minutes. The insoluble dye-polyelectrolyte complex settles out of solution and the absorbance of the supernatant is measured at 520 nm using a 10 cm quartz cell. A calibration curve was developed for each polyelectrolyte and replicates of each sample were measured.

During the initial round of tests, it was determined that residual chlorine interferes the Ponceau dye method. Therefore, sodium thiosulfate was added to each sample that containted chlorine to quench the residual chlorine according to the stoichiometry shown below. As shown, 1.77 mg of sodium thiosulfate is needed per mg of residual chlorine.

$$Na_2SO_3 + Cl_2 + H_2O \rightarrow Na_2SO_4 + 2HCl$$

A battery of tests was conducted to evaluate the potential for sodium thiosulfate to interfere with the Ponceau dye method. The Ponceau dye method was applied to solutions containing various combinations of chlorine and thiosulfate or chlorine, polyelectrolyte, and thiosulfate. There was not a significant difference in the light absorbance at 520 nm due to the presence of concentrations of sodium thiosulfate ranging from 1 to 10 mg/L. Based on these tests, the chlorine interference was counteracted by applying sodium thiosulfate prior to using the Ponceau dye method.

Streaming Current Analysis

Streaming Current Potential has been proposed by many researchers (Bernhardt and Schell 1996; Dentel and Kingery 1989, Dentel et al. 1993, Dentel 1995, Wang 1999) as an alternative to measurement of zeta potential to determine the net charge in a solution and to optimize coagulant dosages. Streaming Current was measured used a Sentrol™ system that contains a transmitter that measures the intensity of the electrical charge and a detector that provides either an absolute or relative measure of the net electrical charge in the water. This system can be operated either on-line or in batch mode. The instrument provides a signal ranging from –5 to 5 volts. In theory, optimum coagulation occurs at a streaming current potential of zero volts. This signal can be correlated to zeta potential (Wang 1999).

Permanganate Reactivity

The reactivity of polyelectrolytes to permanganate was tested by monitoring the rate of permanganate reduction in the presence of known concentrations of polyelectrolytes. Permanganate was added to batch reactors containing polylectrolytes. The residual concentration of permanganate as a function of time was measured using a spectrophotometer.

UV/Persulfate Oxidation

Screening tests were conducted by modification of the methods used for quantification of total organic carbon (TOC). The measurement of TOC is based on evaluation of the mass of carbon dioxide produced from complete oxidation of organic matter under acidic conditions. Depending on the analytical instrument, oxidation reactions are accomplished by combustion, UV oxidation, or oxidation with UV and persulfate. Detection methods for quantifying carbon dioxide are either IR or conductivity, depending on the analytical instrument.

In this study, a Sievers Totoal Organic Carbon analyzer was used for measurement of TOC. The oxidation conditions are programmable to allow for oxidation of each sample by UV alone, UV plus persulfate, or persulfate alone. The UV oxidation reactor is a spiral quartz tube wrapped around a UV lamp. The lamp emits light at 184 and 254 nm. In the absence of persulfate, oxidation is due to the formation of a hydroxyl radical from the photolysis of water:

$$H_2O + h\nu(184\ nm) \rightarrow OH^{\cdot} + H^{\cdot}$$
$$\text{Organic compounds} + OH^{\cdot} \rightarrow CO_2 + H_2O + \text{other}$$

Persulfate can act as an oxidant directly. In the presence of UV radiation, persulfate is oxidized:

$$S_2O_8^{-2} + h\nu\ (184 \text{ and } 254\ nm) \rightarrow 2SO_4^{\cdot-}$$
$$SO_4^{\cdot-} + H_2O \rightarrow HSO_4^{-} + OH^{\cdot}$$

Thus, the coupling of UV and persulfate exposes the sample to a higher dose of hydroxyl radicals, improving the potential for complete oxidation of organic constituents.

While the conditions in the TOC oxidation module do not mimic conditions in a water treatment facility, the operating flexibility of the instrument provides an opportunity to evaluate the amount of carbon dioxide produced from oxidation of each polyelectrolyte due to UV irradiation in the absence and presence of ammonium persulfate. For the screening tests, solutions of each polymer were prepared using Nanopure water to provide concentrations ranging from 0.1 to 5 mg/L. Samples were placed into pre-cleaned 29 x 81 mm 40 mL screw-capped vials with Teflon lined septa and placed into the autosampler. A stainless steel needle was used to transfer samples from the sample vials into the TOC analyzer. The TOC of each dilution was measured under oxidation with UV alone, UV plus persulfate, and persulfate alone. The sample passes through the UV reaction chamber and controlled amounts of 6M phosphoric acid and 15% ammonium persulfate are added. The detection limit of the instrument is 0.05 ppb TOC with a precision of +/-3% of the total carbon. The analysis time for each sample is 6 minutes. Quality assurance protocols included running standards, blanks, and replicate samples. The reagents used for the system are replaced every 90 days.

Concentration of Residual Polymers

The polyelectrolytes used in water treatment practice range in size from about 1,000 to 1,000,000 amu. Typically, the concentration of polyelectrolytes used in water treatment practice ranges from about 0.1 to 5 mg/L. To overcome the problems associated with measurement of low levels of constituents, the feasibility of concentrating residual polyelectrolytes using ultrafiltration was tested. A Pellicon™ Labscale Tangential Flow Filtration (TFF) system was used to concentrate polymers. The system includes a 500 ml acrylic reservoir with pressure gauges, device docking manifold and retentate valve; system base with integral magnetic stirrer and diaphragm pump. The Pellicon™ XL 50 cm^2 membrane is directly docked onto the side of the reservoir, allowing for short tubing connections, thereby minimizing holdup and recirculation volumes.. The membranes tested in this project were composed of regenerated cellulose with polypropylene screens and housing. The membranes were operated with a flux of 7 cm^3/minute at an inlet pressure of 10 psig (0.7 bar). The nominal molecular weight cutoffs for the membranes were 5,000; 100,000; and 10^6 amu.

Prior to using each membrane, it was cleaned with dilute hydrochloric acid (0.1 N) dilute sodium hydroxide (0.1 N) and flushed with Nanopure water until the UV absorbance spectra of each solution was similar to that of Nanopure water. Depending on the pore size of the membrane, the initial cleaning cycle required a minimum of nine repetitions. In some cases it was necessary to pre-heat the sodium hydroxide to 50 ^{o} C. The test procedure was developed to allow for a 10 fold concentration of polyelectrolytes. Stock solutions of polyelectrolytes were prepared at a concentration of 1 g/L (500 mL). Then, each solution was diluted with Nanopure water to yield working solutions of 0.5, and 0.1 mg/L.

Visible and Ultraviolet Absorption Patterns

The use of light absorption to characterize organic materials has been widely applied in characterization of soil and aquatic humic materials. The ratio of absorbance at 465 nm and 665 nm has been termed the E_4/E_6 ratio. It has been reported that the E_4/E_6 ratio is inversely proportional to the molecular weight. A major advantage of this characterization approach is its simplicity and ease of use.

The use of ultraviolet absorption patterns can provide significant insight into the structure of macromolecular organic material. It has been suggested that the ratio of UV absorbance at 253 (A_{253}) to UV absorbance at 203 (A_{203}) be used to indicate dominant functional groups (Korshin et al. 1997). A decrease in the ratio is associated with a decrease in the prevalence of aromatic groups substituted with hydroxyl, carboxyl, ester, and carbonyl groups. A summary of their classification system is given in Table 2.4.

Table 2-4
Summary of functional groups associated with the A253/A203 ratio[a].

A_{253}/A_{203}	Dominant functional groups
>0.4	Carboxylic
.2-.4	Hydroxylic
<0.2	Aliphatic

[a] Adapted from Korshin et al. 1997

Capillary Electrophoresis

Capillary electrophoresis is a separation technique based on the behavior of molecules in an electric field. A literature review was conducted to evaluate the applicability of capillary electrophoresis for quantification of macromolecules. The purpose of the review was to evaluate the types of capillaries, buffers, detectors, and reaction conditions employed for separation of macromolecules. In addition, detection limits and potential interferences were identified. For this project, the methods used for separation of humic materials were adapted for the separation of polyelectrolytes. The reason for taking this approach was to provide a means to discriminate between NOM in a water source and the organic polyelectrolytes. A summary of separation conditions reported for capillary electrophoresis o f humic materials is given in Table 2.5.

Table 2-5
Summary of reaction conditions effective for separation of humic materials using capillary electrophoresis.

Electrophoresis System	Beckman P/ACE 2050	Applied Biosystems	Beckman P/ACE 5510 System	Hewlett Packard
Fused Silica Capillary Size (ID)	50cm x 75 μm	72cm x 50 μm	57cm x 75 μm	56 cm x 75 μm
Detector	UV Photometer	UV Photometer	UV/Visible Diode array	UV/Visible Diode array
Wavelength	214nm	215nm	214nm	214 or 254 nm
Temperature, °C .	300	400	300	300
Voltage	30 KV	15 KV	20KV	15 to 30 kV
pH	8.9	3.17	9	
Buffer	Deionized filtered water with resistivity over 18.3 μohms; 3mM KH_2PO_4; 6mM Na_2BO_7	HCl-L-alanine; 8 mM HCl; 59.8 mM L-alanine	Deionized water with resistivity over 18 μohms; 10 mM borate buffer in 2-propanol;	50 mM acetate, pH: 4.95; 100 mM phosphate, pH 6.95; 90 mM Borate, pH 8.3
Migration time	3 to 7 minutes	4- 20 minutes	3 to 20 minutes	5 to 20 minutes
Type of sample	Humic acids from various locations	Fluka, Aldrich, Janssen	Soil sample	Suwannee River humic and fulvic acids
Concentration	400 mg/: in 1 mM NaOH	5 to 10 mg/L	1 mg/mL in 0.01 M NaOH	> 30 ppb
Reference	Pompe et al. 1996	Rigol et al. 1993	Norden and Dabek-Ziotorzynska 1997	Hozalski and Li 1999

Based on the methods outlined in Table 2.5, a method was developed for separation of polyelectrolytes. Extensive testing was conducted of the use of capillary electrophoresis system to develop a workable method for quantification of polyelectrolytes. A Beckman P/ACE 5000 capillary electrophoresis system was used for the analyses. For this method, a sample is pumped through a small capillary (10-50 μm I.D. , 50-100 cm long) and a potential of 10-30 kV is applied across the capillary to effect separation of charged molecules. The molecules are separated based on their charge/mass ratios and their state of dispersity. The detector is a UV/VIS spectrophotomer for which the capillary diameter is the detection path length. Electrophoretic patterns can indicate different specific binding sites. A major advantage of capillary electrophoresis is its versatility and ease of use. Limited sample preparation is needed and results are obtained fairly rapidly. All tests were conducted at 25 ^{o} C. Three different modes of separation were tested. An overview of the separation principles of each mode is given in Table 2.6.

Table 2-6
Comparison of capillary electrophoresis separation modes

Separation Mode	Definition	Separation Principle
Low pH FSCE	Low pH free solution capillary electrophoresis	Separation of cationic molecules based on mobility
High pH FSCE	High pH free solution capillary electrophoresis	Separation of anionic molecules based on relative mobility
MECC	Micellar electrokinetic capillary chromatography	Use of a migrating micellar phase to separate charged and neutral species based on molecular size.

The capillary used in these tests was a 57 cm uncoated fused-silica CE column (75μm I.D., 300 μm O.D., 50 cm length to the detector). A summary of the conditions tested is given in Table 2.7. The major variables tested were the separation voltage and buffer composition. The pK_a values for the buffers are summarized in Table 2.8.

Table 2-7
Summary of variables tested for capillary electrophoresis

Parameter	Range	Rationale
Wavelength (nm)	200, 214, 254	Optimize sensitivity, detection limits, minimize interferences from buffers
Separation voltage (kV)	-25, -30, -15, -10, +25, +30	Intensity of voltage is related to migration time; polarity of voltage is related to net charge on the molecule
Separation time, min	5, 30, 60	Separation time increases with decreased voltage
Injection time @ 0.5 psi, sec	1, 5, 10, 30	Injection time is proportional to the volume of sample injected.
Buffers	Phosphate; Dextran phosphate, carbonate, borate, acetate, CHES, HEPES	These buffers are widely used in capillary electrophoresis. The pK_a and pH of the buffer influences the net charge on the polyelectrolyte as it passes through the capillary.

Table 2-8
Summary of pK_a values for buffers used in capillary electrophoresis.

Buffer	pK_a	Concentration	Buffer pH
Acetate	4.75	25 mM, 50 mM	4.65
Borate	9.24	25 mM, 50 mM	4.3, 5.3, 6.3, 7.3, 8.3
Carbonate	6.38, 10.25	50 mM	12.5
CHES	9.3	25 mM	Various
HEPES	7.5	25 mM	Various
Phosphate	2.12, 7.21, 12.32	50mM	2.5, 3.5, 4.5, 5.5, 6.5, 7.5, 8.5, 9.5
Dextran phosphate	2.12, 7.21, 12.32	3%, 5%, 10%	2.5, 3.5, 4.5, 5.5, 6.5, 7.5, 8.5, 9.5

Nine buffer systems with different pH ranges were tested to evaluate the effectiveness of polyelectrolyte separation and quantification. The polyelectrolytes tested are listed in Table 2.9. Two of the polyelectrolytes were polyacrylamides and the third one was a polyDADMAC.

Table 2-9
Polyelectrolytes used for method development

Polymer	Net charge	Chemical composition	Concentration (mg/L)
Super floc A-100	Anionic	polyacrylamide	10
Catfloc TL	Cationic	PolyDADMAC	2, 10, 100
Hyperfloc	Cationic	polyacrylamide	2, 10, 100

Size Exclusion Chromatography

Size exclusion chromatography (SEC) is a separation technique based on the hydrodynamic movement of molecules through a porous gel medium. The volume of the gel medium available to molecules is limited by their hydrodynamic volume and thus molecules of different molecular size move through the column at different velocities and can be separated (Kuo and Provder, 1987). Detection is based on using a detector that can discriminate between the molecules and the eluent. For this project, SEC was conducted using a Perkin Elmer 1022 LC with an LC 235 diode array UV detector (separation column : TSK-GEL G5000PW, 05764, 30 cm; 7.5 mm ID, guard column: TSK-GEL PW 05764, 7.7 cm; 7.5 mm ID). The SEC method development tests were conducted using the same polymers shown in Table 2.9 and the concentrations tested were 1, 5, and 10 mg/L of each polymer. The mobile phase was 0.05 M sodium chloride and the flowrate was 0.75 mL per minute. The separation efficiency was tested using individual polymers and combinations of polymers dissolved in distilled-deionized water, tap water, or river water. The effects of oxidant chemical addition on polyacrylamide separation properties were tested on three Superfloc polymers: A-100, A-130, and N-300. Reaction conditions are given in Table 2.10.

Table 2-10
Oxidation Conditions tested for SEC separation of polyelectrolytes

Polyelectrolyte[1] dose, mg/L	Permanganate dose, mg/L	Chlorine dose, mg/L	Oxidation time, hours
5	0	0	1
5	1	0	1
5	0	10	1
5	1	10	1

[1] Polyelectrolytes tested included Superfloc A-100, A-130, and N-300

Bench-Scale Testing of Polyelectrolyte Performance

A standard jar test mixing apparatus was used to conduct parallel bench-scale testing of different chemical dosages. Screening tests were conducted using a simple water matrix with controlled levels of pH, alkalinity, conductivity, and hardness. Inorganic salts were added to deionized distilled water to provide a consistent matrix. For selected tests, turbidity was added to the water in the form of clay particles. Follow-up testing was conducted using natural water sources. Untreated water samples were transported to the University of South Florida for bench-scale testing.

For each series of tests, initial measurements of water quality were taken (pH, alkalinity, UV absorbance, TOC, turbidity). Analytical methods used for water quality testing are outlined in Table 2.11. The typical procedure for the jar testing was developed to mimic a conventional water treatment facility. All tests were conducted using a Phipps and Bird Jar test apparatus consisting of six square 2L reactors and six agitator blades mechanically coupled to a variable speed drive. Coagulant chemicals and polyelectrolytes were added to each reactor under conditions of rapid mixing. Then the mixing speed was reduced to simulate a three stage decayed gradient flocculation period. After flocculation, a 60 minute quiescent period was used to simulate sedimentation.

Water quality analyses were conducted prior to the test and at intervals during the test. Samples were withdrawn for analysis of turbidity, UV absorbance (254 and 203 nm), TOC, alkalinity, hardness, and pH. For tests containing oxidants (permanganate or chlorine), the concentration of residual oxidant was measured and the rate of oxidant consumption was monitored. Tests were also conducted to measure the filterability of residual particles. Selected samples were processed for analysis of disinfection byproduct formation potential and polymer characteristics.

Polyelectrolytes were paired with other chemicals likely to be present concurrently in the treatment system. Polymers were obtained from water treatment utilities. Tests were quantified using end-points for polymer characterization. The primary focus of this effort was on oxidant chemical interactions.

Table 2-11
Analytical methodology.

Parameter	Standard Method	Instrument	Detection Limits
Alkalinity	2320	Titration	10 mg/L as $CaCO_3$
Chlorine total and free	4500-Cl D Amperometric Method	Fisher Scientific CL Titrimeter 397	0.01 mg/L as Cl_2
Conductivity	2510B	Orion Model 135 Conductivity meter and probe	1 μS/cm
Particle size and number	2560 D	Spectrex Laser Particle Counter Model PC-2000	1 particle /mL 0.5 to 100 μm
pH	4500-H^+ B Electrometric method	Fisher Scientific Accumet Basic pH meter and probe	0.1 pH unit
PolyDADMAC	Colorimetric Method (Ponceau S Dye)	Hach DR 4000U; Absorbance @ 520 nm	0.5 mg/L
TOC	5310 C Persulfate-Ultraviolet	Sievers TOC Analyzer Model 800 with Autosampler	10 ppb
Turbidity	2130	Hach Laboratory Turbidimeter Model 2100AN	0.001 NTU

Polymer Effectiveness Testing

Changes in the effectiveness of polylectrolytes due to reactions with other chemicals may result in modified polymer reactivity towards suspended and colloidal particles. Tests were developed to quantify the relative effectiveness of polymers in the presence or absence of other chemicals under standardized conditions. A standardized particle suspension was used in each test. The particle size distribution and rate of settling of the particle suspension were evaluated after reacting with a fixed concentration of polyelectrolyte.

The focus of the testing was on evaluation of the impacts of chlorine, permanganate, and ultraviolet light on polyelectrolyte performance in coagulation/flocculation of a Kaolinite suspension. Each polyelectrolyte was tested directly and following exposure to oxidant chemicals and/or ultraviolet light. The influence of oxidants on polyelectrolyte performance was tested by adding a fixed amount of either chlorine (5 mg/L) or permanganate (0.8 mg/L) to each reactor. The combined effects of chlorine and ultraviolet light were also evaluated. The test conditions are summarized in Table 2.12.

Table 2-12
Summary of test conditions used for polyelectrolyte performance testing

Parameter	Description
Polyelectrolytes	Catfloc TL, Superfloc N-300
Dosage range	0 to 1 mg/L of each polyelectrolyte
Oxidant chemicals	Potassium permanganate, 0.8 mg/L; Chlorine, 5 mg/L
UV irradiation	Irradiate polyelectrolyte prior to addition to each reactor by exposing solution to UV light at 254 nm for 30 minutes.

A stock Kaolinite suspension was prepared by equilibrating Kaolinite in a 0.3M solution of sodium hydroxide at 105 °C. The Kaolinite was then suspended in a 1 M salt solution (sodium chloride). After an initial mixing period, the pH was adjusted to about 7.0 and the suspension was maintained under quiescent conditions for 7 days to allow for particle sedimentation and stabilization. The turbidity of the stabilized suspension was about 2000 NTU per mL. The properties of the test suspension are given in Table 2.13 and the particle size distribution of the suspension is given in Figure 2.10. For each polyelectrolyte, stock solutions were prepared at a concentration of 1 mg/mL. For tests involving UV irradiation, the polyelectrolyte was irradiated prior to injection into the reactors. A syringe was used to add the polyelectrolyte to each reactor.

Table 2-13
Properties of Kaolinite suspension used for evaluation of polyelectrolyte performance

Parameter	Value
pH	7.4
Alkalinity	100 mg/L as $CaCO_3$
Conductivity, μS/cm	365
Hardness	100 mg/L as $CaCO_3$
Turbidity	2 NTU

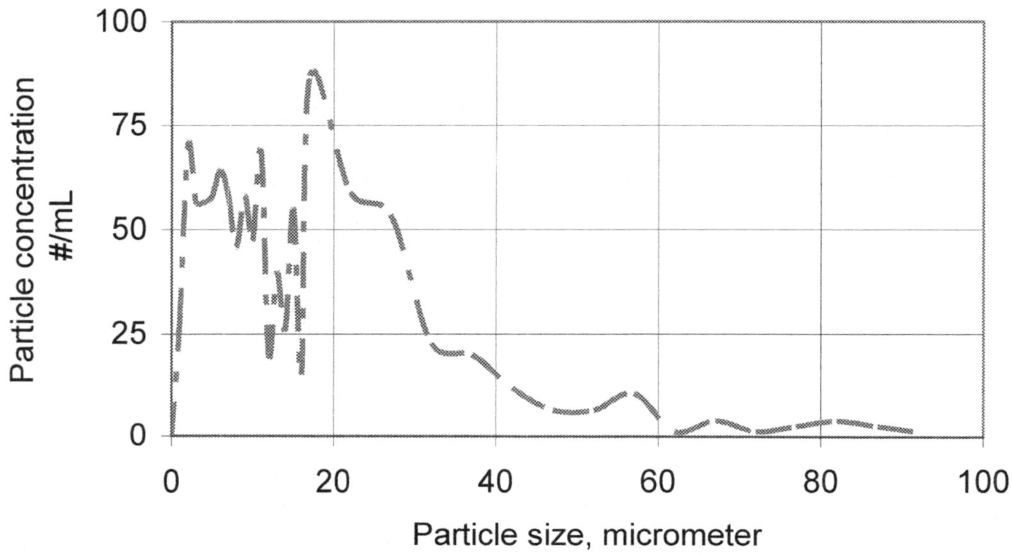

Figure 2-10. Example particle size distribution of Kaolinite suspension with turbidity of 2 NTU, pH 7.4, and alkalinity and hardness of 100 mg/L as $CaCO_3$.

The polymer effectiveness testing was conducted using a jar test approach. Six parallel square batch reactors were used to test polyelectrolyte performance. For each reactor (jar), 1 mL of the stock Kaolinite suspension was added to 1 liter of Nanopure water amended with 100 mg/L of calcium carbonate and adjusted to pH 7.4. The residual pH, alkalinity, hardness, turbidity, particle size distribution, settling velocity, streaming current, UV spectra, and total organic carbon were measured on each sample, In addition, a 100 mL sample was withdrawn from each reactor and gravity filtered through a Whatman #40 filter. The time to filter 100 mL was recorded as a measure of suspension filterability.

Ultraviolet light reactions

Several protocols were developed for evaluating the degree to which UV irradiation influences the reactivity of organic polyelectrolytes. A summary of the tests is given in Table 2.14. Initially each polyelectrolyte was suspended in Nanopure water to test reactivity independent of water matrix impacts. Polyelectrolytes that displayed reactivity through these tests were subjected to more comprehensive testing in simple water matrices.

Table 2-14
Summary of UV irradiation screening tests.

Test	Approach
Evaluation of TOC recovery	Measure TOC of polyelectrolytes under different oxidation conditions Evaluate carbon dioxide production from UV oxidation Evaluate carbon dioxide production from UV plus persulfate oxidation
Evaluation of the stability of UV spectra	Expose samples to UV irradiation in a UV spectrophotometer. Compare variations in UV spectra over time
Evaluation of controlled UV exposure	Use a UV reaction chamber and expose samples to UV light for fixed periods of time. Evaluate characteristics of polyelectrolytes after UV exposure
Evaluation of UV exposure coupled with other oxidants	Apply all of the above techniques following pre-treatment with chlorine and potassium permanganate
DBP formation	Evaluate DBP formation potential from polyelectrolytes that exhibit positive responses to other tests
Sunlight exposure	Simulate UV exposure from sunlight, Expose water containing polyelectrolytes to sunlight in the absence or presence of chlorine or permanganate, Evaluate impacts on polyelectrolyte characteristics and DBP formation potential

Two approaches were used for exposing the polymer solutions to ultraviolet light: sunlight or a UV sterlizer. For each sunlight exposure test, the polymer was mixed with Nanopure water using glass beakers that had been acid washed and oven dried. The nominal concentration of polymer in each sample was 5 mg/L. A baseline alkalinity of 100 mg/L as $CaCO_3$ was used. Sodium hypochlorite was added to selected reactors at a dose of 2 mg/L as Cl_2. Initially, the pH, alkalinity, chlorine, and polymer concentration of each sample were recorded. Samples were placed in direct sunlight on the roof of the Kopp Engineering building (University of South Florida, Tampa) for 30 minutes under conditions of high ultraviolet radiation (Noon to 2:30 pm, clear sky, >80° F). The depth of sample in each reactor was 6 cm. Following exposure to sunlight, samples were covered and water quality analyses were conducted.

As an alternative to sunlight, a Millipore UV sterilizer was used as source of ultraviolet radiation. The test unit provides continuous exposure to UV irradiation at a fixed wavelength of 253.7 nm in a closed chamber. Each unit contains 4 UV lamps that operate at low wattage. Samples of fixed concentrations of polyelectrolytes were set-up in Petrie dishes and exposed to UV radiation in the presence or absence of other water treatment chemicals for controlled time periods. Following exposure to UV, the properties of the polyelectrolyte were quantified to determine if there are any measureable impacts from UV exposure.

Performance Measures of Polyelectrolyte Effectiveness

In consort with the laboratory testing, a set of performance measures were used to provide a basis for comparing different chemical interactions. Ideally, the use of polyelectrolytes for turbidity removal should produce a measurable change in the rate of change of particle size and a net change in the zeta potential of the solution. The silt density index may provide a basis for evaluation of the relative effectiveness of different polyelectrolytes on a fixed particle suspension. A summary of the performance indicators used for this project is given in Table 2.15. To supplement laboratory testing, operating data was obtained from each of the participating utilities.

Table 2-15
Summary of performance indicators under development for this project

Process	Performance Indicator
Coagulation, Flocculation	Reaction rate, rate of change of particle size, rate of TOC conversion, mixing intensity, zeta potential
Sedimentation	Particle settling velocity, particle size distribution, zeta potential, settled water quality (turbidity, color, TOC), Silt density index
Filtration	Particle removal efficiency, headloss development, comparison with in-plant measurements
Filer backwash	Water quality, residual polyelectrolytes, particle size distribution, particle settling velocity

Disinfection By-Product Formation Potential Testing

Analysis of disinfection by-product formation potential was based on evaluating the production of trihalomethanes (THMs) and Haloacetic acids (HAAs) in response to exposure to a fixed dosage of chlorine for a fixed incubation period. Chlorine was added to each sample at a concentration of 6 mg/L using 300 mL batch reactors (BOD bottles with ground glass stoppers). The chlorine demand of each sample was determined after a 30 minute contact time. Each sample was incubated in the dark at a constant temperature (20°C) for 7 days. At the end of the incubation period, samples were transferred into 40 mL pre-cleaned amber glass vials containing a fixed amount of sodium thiosulfate to quench the residual chlorine. Residual oxidant, chlorine (total and free), UV, TOC, and alkalinity were measured in each sample after the incubation period. Analyses for THMs and HAAs were conducted by the City of Tampa Water Department.

Filter Media Interactions

An experimental study was conducted by the Australian Research team (CSIRO) to determine the effect of precoating filter media with various polyelectrolytes on filtration efficiency. This method of immobilizing polyelectrolyte was intended to provide a reactive surface on the filter media to allow for efficient use of polyelectrolytes.

Tests were conducted using a pilot scale version of a conventional water treatment plant. The Continuous Pilot Scale Treatment Plant (CPSTP) is based on the standard jar test but allows for continuous flow of water through the series of tanks. A diagram of the CPSTP is shown in Figure 2.11. The system was operated at a flowrate of about 0.5 L/min. Under continuous operation, the untreated water was pumped into the coagulation chamber ('1') using a throttled centrifugal pump. The mixing speed in the contact vessel was about 160 rpm. Alum was dosed at about 40 mg/L (based on jar tests). Water then flowed to the flocculation chamber and was stirred at 30 rpm. Polyelectrolytes were added to the flocculation chamber and particle removal occurred in the inclined sedimentation unit. Filtration was used after sedimentation. The sand used in the filter was a finer grade than is used in standard filters (16/30). The media depth was about 40 cm. The source water for this project was derived from a permanent creek outside the Melbourne, Australia metropolitan area. The water was high in color (~100 PCU) and had turbidity values below 10 NTU. The turbidity of the clarified water ranged from 6-10 NTU, resulting in short filter run times.

Initially, tests were conducted to optimize alum dose, filter depth, filter throughput, and filter run time. The sand particles were contacted with different doses of flocculants external to the pilot plant and then filter runs under different sets of operating conditions (ie different alum doses and/or different pH values and with and without polyelectrolyte aids) were conducted. Filtration and back-washing efficiencies were monitored by on-line measurement of turbidity in the filtered waters and backwash liquors respectively.

For the pre-coating studies, approximately 50 g sand (16-30 mesh filter sand from the Comadai deposit, Victoria, Australia) was stirred at 1000 rpm in 100 ml water in a 250 mL beaker for 1 minute before the liquor was decanted. This was repeated 20 times. The final wash water from each sand was kept as the blank for TOC measurements. Polyelectrolyte solutions were prepared at 0.1%. The 50 g washed sand was suspended in 100 mL water by stirring at 1000 rpm then polyelectrolyte was added (0.5 mL, 5 mL, 50 mL, or 500 mL) as required for the loadings selected). After 5 min the stirring was stopped and the sand settled. Supernatant was removed for analysis and the sand was drained. The characteristics of the polyelectrolytes used to pre-coat the sand are summarized in Table 2.16. The chemicals were selected to represent a range of charge and molecular weight properties. The sand was characterized before and after each set of testes. Sand was removed from the column after a filter run using a piece of 15 mm rigid plastic hollow tubing which was tapped through the sand bed. The core of sand in the tubing was tapped out in three sections. Sand samples after backwashing were taken when the sand was removed from the column.

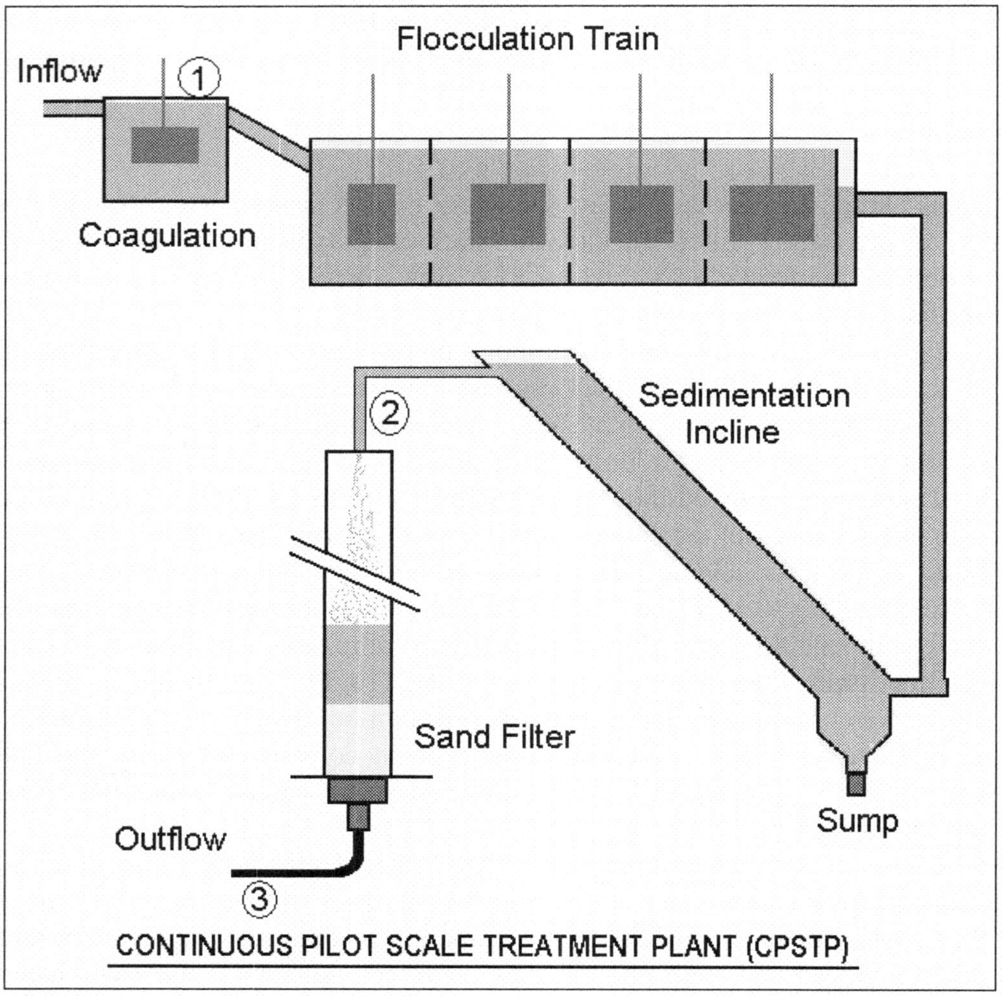

Figure 2-11. Schematic of Continuous Pilot Scale Treatment Plant Used for testing.

Table 2-16
Comparison of Chemicals Used in Polyelectrolyte Testing

Polyelectrolyte	Charge	Molecular Weight
Magnafloc E10	10% anionic	Very high
Magnafloc LT20	Neutral	High
Magnafloc LT35	100% cationic	Medium
Betz 1160x	Moderately cationic	High

Quantification of polyelectrolyte adsorption on filter media

Several approaches were evaluated for monitoring the extent of polyelectrolyte adsorption on filter media. The use of dissolved organic carbon measurements, viscosity, and streaming current measurements was evaluated.

Dissolved Organic Carbon. Initially, the adsorption of the polyelectrolyte on sand was evaluated by measuring the decrease in dissolved organic carbon (DOC) in the supernatant. Solutions were analyzed for DOC using an O-I-Analytical Model 1010 Total Organic Carbon Analyzer with automatic sampler. Solutions were diluted if necessary and vacuum filtered through 0.45 μm sterilized membranes (Gelman) prior to DOC analysis. Polymer concentrations ranged from 0.005 mg/mL to 0.83 mg/mL. The membrane filters became clogged at the higher concentrations and it was necessary to dilute some solutions before filtration, introducing significant sources of error. Experimentally it was found that up to 20% of polymer was retained on the filter membrane, particularly at higher polymer dosages. This was also evidenced by the presence of a sticky residue on the filter membranes after filtration. Therefore, the use of DOC was not reliable for quantifying the extent of polyelectrolyte adsorption.

Viscosity. A Brookfield DV11+ viscometer was used to measure the viscosity of solutions containing various concentrations of polyelectrolytes. For high molecular weight polymers, there was a linear relationship between viscosity and concentration. A calibration curve was developed and the viscosity of the polyelectrolyte solution was evaluated before and after contact with the sand.

Streaming Potential Measurement. Streaming potential measurements were performed on untreated and polymer treated sands. In each case, the sand/polymer slurry was mixed for 10 minutes at 1000 rpm. The liquid was drained and the sand was washed with water twice. A capillary tube of around 12 cm in length, 2 mm diameter, was filled with the wet sand for streaming potential measurements. Each end of the capillary was capped with a small fine mesh membrane housed in a plastic casing. Each casing had a small electrode that was connected by alligator clips to a voltmeter. 10^{-3} M NaCl was placed into a continuously stirred insulated reaction cell that was connected to the sand capillary. The flow direction was controlled by a pump. The conductivity was about 130 μS/cm and the pH of the NaCl solution was about 6. The pressure of the NaCl flow was measured with a connection at the positive pressure end in these experiments. The pressure and potential of the capillary was measured over a 30 sec interval and the data were converted to a value for the streaming potential.

The presence of air bubbles in the sand capillary were a constant problem, thus in all experiments nitrogen gas was bubbled through the NaCl solution at a pressure of around 100 kPa. Typically before each measurement, the liquid was pumped through the system for 10-20 minutes each in both directions. Measurements of potential were recorded with the NaCl flowing into the cell at the pressure measurement end.

Conduct of filtration tests

The source water for the filtration tests was obtained from Aura Vale Lake. For the first few tests, the turbidity, color and pH were fairly consistent. However, due to heavy rain, the turbidity of water collected for the LT35 run rose from around 2 NTU to 7 NTU and remained at this level for the remainder of the experiments. This affected the breakthrough time for untreated sand. At a turbidity of around 2 NTU, the breakthrough time was around 6 hours; while for water with a turbidity of 7 NTU this time decreased to around 3.5 hours. It is interesting to note that the optimum alum dose as determined by jar tests was unchanged by the turbidity change.

The filtration tests were designed to test each polymer under different surface concentrations. A summary of the filtration tests is given in Table 2.17. A series of up to 3 runs were conducted for each sand preparation. The sands were kept wet until needed for streaming potential measurements. Each experiment was operated until breakthrough was achieved and the turbidity reached at least 1 NTU. After each experiment, a backwash was performed on the sand filter and the turbidity was monitored over time. The washings were kept along with the settled flocs from the clarifier. Mass balances were performed on each sample.

Table 2-17
Summary of test conditions for polyelectrolyte coating of filter media

Filter sand conditioning	Polyelectrolyte dose, mg/g sand	Alum dose, mg/L	Water Quality		Number of runs
			Turbidity, NTU	Color, PCU	
Natural sand	0	60	2.1	47	4
Natural sand	0	60	7.1	51	3
Polyelectrolyte					
Betz 1160x	1	60	3.3	46	2
Betz 1160x	0.1	60	8.0	57	3
Betz 1160x	0.01	60	7.2	61	1
Betz 1160x	0.05	60	7.2	61	2
FITC polyacrylamide	0.05	60	6.2	67	1
FITC polyDADMAC	0.05	60	6.2	67	1
Magnafloc E10	1	60	2.0	48	3
Magnafloc E10	0.1	60	7.2	61	2
Magnafloc E10	0.05	60	7.2	61	2
Magnafloc LT35	1	60	7.1	44	3
Magnafloc LT35	0.05	60	6.2	67	2

For each filter run, approximately 1700 mL sand was fluidized with tap water for 10 minutes to remove contaminants. After washing, the sand was 'tapped' to give a final tapped volume of 1600 mL of clean sand. This sand was added to a mixing vessel and enough polymer was added to produce a ratio of approximately 1 mg polymer/g sand. The sand/polymer slurry was mixed using an overhead stirrer at around 300 rpm with a large paddle stirrer for 10 minutes. After this time, a sample of supernatant was taken for viscosity measurements (along with some original polymer solution to determine the amount adsorbed to the sand). Typically around 0.2 mg of polymer /g sand was adsorbed by the sand for Magnafloc E10 and 0.6 mg polymer/g sand for BETZ1160x at a concentration of 1 mg /g sand. Viscosity measurements were not possible for LT35 due to the less viscous nature of this polyelectrolyte. After rinsing, the liquid was decanted and another rinse repeated.

The sand was then "washed" into the pilot plant's 2.7 m column. The flocculator and clarifier sections of the pilot plant were started up for around 30-60 minutes for flocculation equilibration prior to connecting the clarified water to the column. The turbidity was monitored on-line and hourly grab samples were collected for color, pH and column water height.

Adsorption of Tagged Polymer

Another approach that was tested to determine the fate of adsorbed polyelectrolyte was the use of fluorescently tagged polyelectrolytes prepared in the CSIRO laboratories. Experimentally it was necessary to find conditions for measurement of the fluorescently tagged polyelectrolyte that could be used for samples of both product water and backwash water. This was accomplished by carrying out all fluorescence measurements with the pH adjusted to around 11.5 in 2 M KCl. Firstly, flocs were dissolved to release of attached flocculant before measurement. This was done by adding about two drops of concentrated NaOH to the floc suspension. Measurements were done in 2 M KCl to reduce quenching of the fluorescence from the tag by the added NOM. It was found to be necessary to control the pH accurately to 11.5 ± 0.2 as the fluorescence was found to vary with pH. Excitation was carried out at 493 nm with 2.5 nm slit; emission was scanned from 500 to 550 nm with 4 nm slit. The instrument used was a Perkin Elmer LS50B Luminescence Spectrometer with 1cm x 1cm cell.

Fluorescent polyelectrolyte calibration curve.

A stock solution of 0.0136 g tagged polyelectrolyte was made up in 100 mL. A series of 50 mL standards were then prepared in the range of 0 to 3 ppm. Potassium chloride (7.455g) was added to each solution to bring the concentration to 2 M. The pH of each standard was adjusted to pH 11.5. The calibration curve is shown in Figure 2.12.

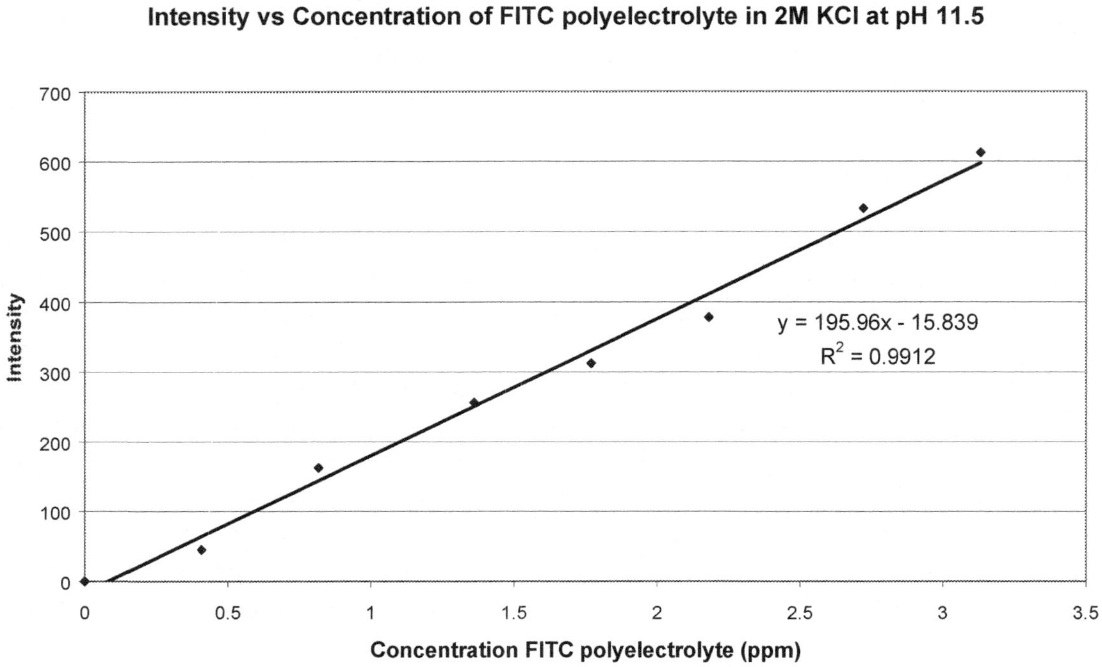

Figure 2-12 Calibration curve for FITC tagged polyelectrolyte.

Fluorescent polyelectrolyte with added NOM – calibration curve.

A series of standards were prepared by keeping the polyelectrolyte constant at 3 mg/L for each standard and adding varying amounts of NOM. A batch of backwash collected from a pilot plant run with untreated sand was used to dose the tagged polyelectrolyte with NOM. This sample had a solids content of 0.0856 g solids/200 mL. After spiking tagged polyelectrolyte with the backwash flocs the mixture was adjusted to pH 11.5 to dissolve the flocs, 7.455 g of KCl was added and made up to 50 mL. The color of each standard above was measured by the spectrophotometer at 400 nm against a 50 PCU standard. Each standard was filtered through a 0.45 µm Gelman Sciences filter membrane prior to measurement. In most cases it was necessary to dilute the standards to bring them in the color range on the calibration curve. The calibration curve is shown in Figure 2.13.

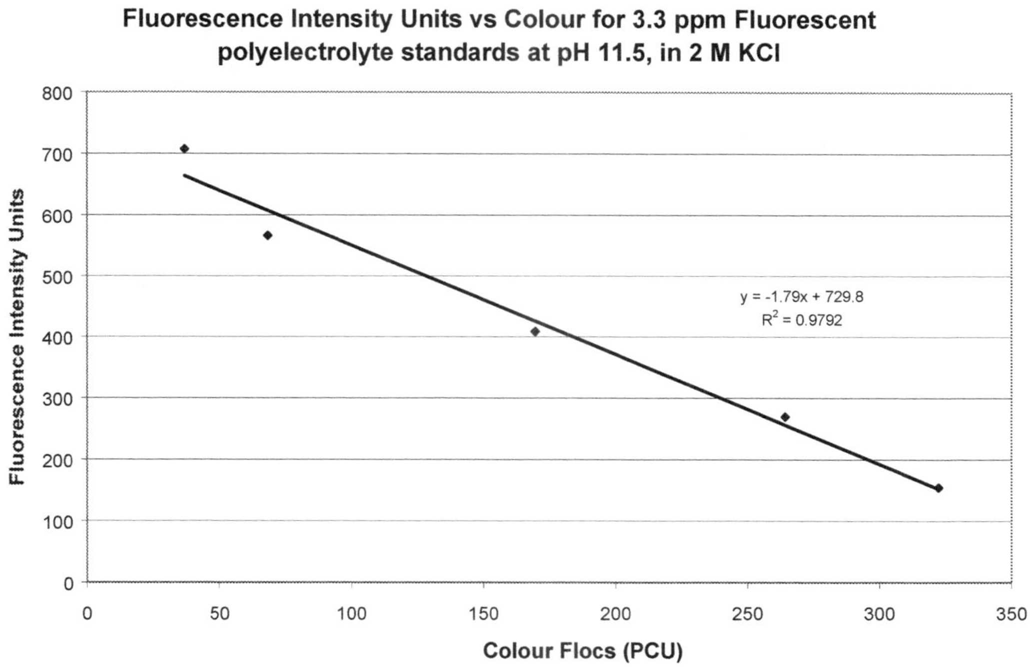

Figure 2-13. Effect of Color on Fluorescence Reading.

CHAPTER 3: RESULTS AND DISCUSSION

This project encompasses a variety of approaches to evaluate the extent to which chemical interactions occur between organic polyelectrolytes and other chemicals used for water treatment applications. The results are presented in three sections: assessment of chemical use patterns, results of bench-scale testing, and results of filtration precoat testing.

CHEMICAL USE PRACTICES

Several sources of information were used to evaluate chemical use practices in the water industry. An overall summary of data obtained from the AWWA database,WATER:\STATS, is presented in Table 3.1. As shown, a higher proportion of surface water treatment facilities use organic polyelectrolytes as compared to groundwater treatment facilities. Anionic polyelectrolytes tend to be the dominant polymer used in groundwater treatment plants whereas cationic polymers are used more extensively in surface water treatment plants

Table 3-1
Summary of chemical use practices reported in the WATER:\STATS database

Water source	Groundwater	Surface water
Number of plants surveyed	492	542
Percent of surveyed plants using polymer	13	66
Anionic polymer	42	7
Cationic polymer	25	48
Nonionic polymer	11	13
Use of multiple polymers	0	10

Information available on water utility websites was surveyed to determine the typical sequence of chemical addition. Since utility websites tend to provide different types of information in a variety of formats, consistent information was not available for all facilities. In addition, increased security measures implemented after 9-11-01, have impacted the availability of information. A summary of chemical sequence information is given in Table 3.2. As shown, inorganic coagulants were the first chemical added in over half of the facilities evaluated. Oxidants were the first chemical added in about 22 percent of the facilities. This information was used to help design the survey that was developed for this project.

Table 3-2
Summary of sequence of chemical use reported on water utility websites.

Chemical	Initial chemical, % of facilities	Second chemical, % of facilities	Third chemical, % of facilities
Alum	43	0	14
Ferric chloride or ferric sulfate	9	13	9
Oxidant (chlorine, permanganate, or ozone)	22	8	36
Polymer	8	9	4
Other	18	62	41
None reported	0	8	10

Survey Results

The majority of the participants in the On-Line Survey responded through e-mail requests. Over 500 e-mail letters were sent to utilities throughout the US. No effort was made to pre-screen the utilities in terms of size, location, source water or ownership. The goal was to solicit participation from as many utilities as possible and to try to obtain information from throughout the U.S.. About 10% of the e-mail requests were rejected due to either incomplete contact information or security concerns. The response rate for the remaining requests was about 20%. Of the total responses, 84% were from publicly owned utilities. About 68% of the utilities use a single type of source was whereas the remaining utilities rely on a mixture of sources that vary seasonally.

The distribution of water sources associated with the utilities that responded to the survey is summarized in Figure 3.1. As shown, the majority of the responses were associated with surface water sources. Wholesale sources represent a combination of surface and ground water sources, depending on the wholesale provider, location, and time of year. The geographical distribution of the survey respondents is shown in Table 3.3. While utilities in all states were invited to respond, responses were received by utilities in 30 states that ranged in capacity from about 2 MGD to over 100 MGD. As would be expected, the distribution of water sources varied by location.

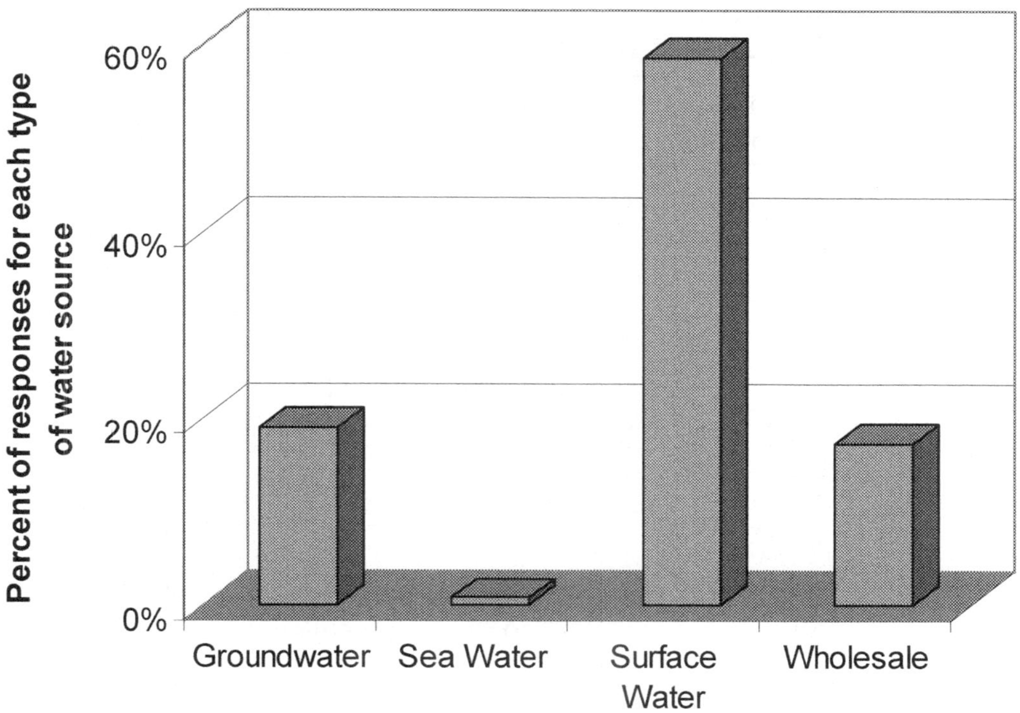

Figure 3-1. Distribution of types of source water associated with the utilities that participated in the On-Line Water Treatment Survey.

Table 3-3.
Geographical distribution of utilities that responded to the On-Line Water Treatment Survey

State	Percent of total responses	Treatment capacity, MGD	% ground water	% surface water	% wholesale
Alaska	1.2	3 to 5	100	0	0
Arizona	2.5	2 to over 100	0	100	0
California	11.1	5 to over 100	16	53	26
Colorado	5.0	3 to over 100	0	67	33
Connecticut	5.0	2 to over 100	11	89	0
Florida	7.4	20 to over 100	67	33	0
Georgia	1.2	75 to over 100	0	100	0
Illinois	1.2	Over 100	0	100	0
Indiana	6.2	5 to 50	80	20	0
Iowa	2.5	5 to 20	50	50	0
Kentucky	1.2	5 to 10	0	100	0
Louisiana	1.2	20 to 50	100	0	0
Maine	1.2	50 to 75	0	100	0
Maryland	3.7	10 to over 100	20	80	0
Massachusetts	1.2	20 to 50	0	100	0
Michigan	7.4	10 to 50	0	57	43
Missouri	3.7	20 to over 100	17	83	0
Nebraska	2.5	50 to 75	33	67	0
New Hampshire	1.2	20 to 50	25	75	0
New Jersey	1.2	50 to 75	0	100	0
New York	2.5	20 to 50	50	50	0
North Carolina	3.7	1 to 75	25	75	0
Ohio	2.5	50 to over 100	0	100	0
Pennsylvania	1.2	over 100	0	100	0
Texas	5.0	20 to over 100	13	75	12
Utah	2.5	20 to over 100	0	100	0
Virginia	7.4	3 to over 100	29	14	57
Washington	3.7	10 to over 100	50	50	0
West Virginia	1.2	2 to 3	0	100	0
Wisconsin	2.5	3 to over 100	0	100	0

Comparison of the Types of Polyelectrolytes in Use in Water Treatment Facilities

About 15% of the utilities that responded to the survey use polyelectrolytes for treatment. In surface water treatment facilities, the polyelectrolytes are used for coagulation, flocculation, ballasted sedimentation, filtration, and filter backwash treatment. In groundwater treatment, polyelectrolytes are used in conjunction with lime softening to function as a coagulant. All of the facilities that use polyelectrolytes reported that they use "finished" water for make-up water. While limited information was provided on the chlorine levels of the make-up water, there is a strong potential for chlorine to react with the polyelectrolytes prior to chemical addition.

Coagulation/flocculation

The dominant use of polyelectrolytes reported in the survey is for coagulation and flocculation. A summary of the types of polyelectrolytes reported to be used at different facilities is given in Table 3.4 for coagulation applications. The majority of the polyelectrolytes used for coagulation are cationic polyDADMACs. Anionic and nonionic polyacrylamides are also used. There did not seem to be any strong water quality trends associated with the use of polyDADMACs as compared to polyacrylamides.

Table 3-4.
Distribution of Polyelectrolyte use for Coagulation from Responses to On-Line Water Treatment SurveyCoagulation

Parameter	Type of polyelectrolyte		
	Polyacrylamide	PolyDADMAC	Unknown
Percent of responses	7	49	14
Vendors	Sweetwater, Stockhousen, Cytec	Polydyne, Nalco	
Source water turbidity range, NTU	2 to 3	2-10	1 to 2
Source water TOC range, mg/L	0.5 to 1	3 to 4	3 to 4
Charge			
Anionic	67		
Cationic		100	100
Nonionic	33		
Charge Density	0.5 to 1	50 to 100	
Molecular Weight Range	<1,000 to 100,000	1,000 to 10^6	

In contrast to patterns observed for coagulation polyelectrolytes, the majority of the polyelectrolytes reported to be used as flocculation aids are either anionic or nonionic polyacrylamides. PolyDADMACs are also used by some utilities as flocculation aids. A summary of reported polyelectrolyte use for flocculation is shown in Table 3.5. Again, there were no strong trends that related water quality to the use of either polyacrylamides or polyDADMACs.

Table 3-5.
Distribution of Polyelectrolyte use for Flocculation from Responses to On-Line Water Treatment Survey

Parameter	Type of Polyelectrolyte	
	Polyacrylamide	PolyDADMAC
Percent of responses	80	20
Vendors	Nalco, Ciba, Cytec	Sweetwater
Source water turbidity range	1 to 20	
Source water TOC range	0.5 to 5	3 to 4
Charge		
Anionic	50	
Cationic		100
Nonionic	50	
Charge Density	10 to 50	50 to 100
Molecular Weight Range	500,000 to 10^6	

Filtration

A summary of reported use of polyelectrolytes as filter aids is given in Table 3.6. In general, filtration aids were either nonionic or anionic polyacrylamides. Limited information was provided on the characteristics of the polyacrylamides (molecular weight, charge density). Similar types of polyacrylamides are used for filter backwash treatment.

Table 3-6.
Distribution of Polyelectrolyte use for Filtration from Responses to On-Line Water Treatment Survey

Parameter	Type of Polyacrylamide	
	Nonionic	Anionic
Filtration aids		
Percent of responses	50	50
Vendors	Cytec, Polydyne	Cytec
Filter backwash treatment		
Percent of responses		100
Vendors		Cytec, Polydyne

Use of multiple chemicals

The survey database was used to identify utilities that add multiple chemicals through treatment and to compare the sequence of chemical addition. A summary of data from the survey for cases where polyelectrolytes are used in tandem with other chemicals is given in Table 3.7. The stages of treatment are defined based on conventional surface water treatment with pre-oxidation at the headworks of the treatment plant or at the source water intake. In some cases oxidants are added upstream of polyelectrolytes. For the utilities that responded to the survey, the pre-oxidants included chlorine dioxide, ozone, or potassium permanganate. Thus, depending on the reaction rate of the pre-oxidant, there is a possibility that the oxidant chemicals are available to react with polyelectrolytes that are added during coagulation and flocculation. Chlorine or chloramines are often added either before or after filtration. Residual polyelectrolytes could potentially react with chlorine or chloramines. The majority of the facilities that responded to the survey had treatment systems that were exposed to the environment. Depending on the location, there is a potential for sunlight (UV) interaction with residual polyelectrolytes and oxidant chemicals.

Summary of survey results

The survey was designed to compile information that could be used to assess water quality and operational variables that influence the selection and performance of polyelectrolytes. It was also designed to identify the types of chemicals that are likely to be used concurrently with polyelectrolytes. It is a valuable tool, however, the results of the survey are limited by the relatively small dataset of responses. It would be useful to figure out a way to encourage more utilities to respond to the survey.

Table 3-7.
Types of chemicals reported to be used in facilities that use polyelectrolytes[1]

Treatment step where chemicals are added						
Pre-oxidation	Coagulation	Flocculation	Filtration	Post-filtration	Filter backwash	Sunlight
	Alum	Sweetwater 10		Chlorine		Yes
Chlorine dioxide	Alum	Nalclear 8181		Chlorine; Sodium silicate		Yes
Ozone	Clarifloc 318; Ferric sulfate	Chloramines				Yes
Powdered Activated Carbon	C308 P Ferric chloride	Chlorine; Ferric chloride	Ferric Chloride			
Potassium permanganate	Cytec 572C; Alum		Cytec 1849A	Chlorine:; Sodium hexametaphosphate	Cytec 1849A	Yes
Potassium permanganate	CatFloc TL; Ferric chloride	Superfloc A100	Chlorine; Superfloc A100	Zinc orthophosphate		No
Polyphosphate	Lime (softening); Praestol 2515T		Chlorine			Yes

[1]Data derived from On-Line Water Treatment Survey

BENCH SCALE TESTING OF POLYELECTROLYTE PERFORMANCE

An array of tests were conducted to evaluate polyelectrolyte performance. First analytical procedures were developed and verified to allow for monitoring of bench-scale tests. Results of analytical testing, bench-scale performance tests, and ultraviolet exposure tests are presented in this section.

Quantification of Polyelectrolytes

Quantification of the concentration of unreacted polyelectrolytes and transformed polyelectrolytes is complicated by the low concentrations and potential interferences from other constituents in water. The analytical procedures used in this project included membrane concentration of residual polymers, indirect and direct tests for quantifying residual polymers Indirect tests included dye complexation, streaming current measurements, organic carbon analysis, and disinfection byproduct formation potential tests.. Direct tests included capillary electrophoresis and HPLC.

Concentration of residual polymers

Ultrafiltration was used to concentrate residual polymer and the results of the concentration step were evaluated using a mass balance approach. A summary of the ultrafiltration conditions tested is given in Table 3.8. After each round of sample concentration, the membrane was subjected to intensive cleaning until all residual polyelectrolyte was flushed from the system.

Table 3-8
Summary of ultrafiltration conditions for concentration of polyelectrolytes

Polyelectrolyte Concentration	Membrane pore size	Initial Volume	Final Volume	Verification
0.1 mg/L	5000 amu	500 mL	250 mL	Ponceau dye, UV, TOC
0.5 mg/L	5000 amu	500 mL	100 mL	Ponceau dye, UV, TOC

Indirect tests for quantification of polyelectrolyte concentration

Indirect tests for quantification of residual polyelectrolytes included dye complexation, streaming current measurements, organic carbon analysis. In the absence of other sources of organic matter (NOM, etc), indirect tests provide a measure of the mass of polyelectrolyte in solution, but do not provide insight into molecular changes that may occur as a result of chemical reactions (oxidation, reduction, hydrolysis, etc.).

Dye Complexation. For cationic polyelectrolytes, residual concentrations can be evaluated using a colorimetric method (Fielding et al., 1999, Parazac et al., 1987). In this method, cationic polyelectrolytes are reacted with an anionic dye to form an insoluble complex. The complex precipitates from the solution and the amount of dye remaining is measured spectrophotometrically. In principle, the dye reacts with all macromolecular cationic components of the solution. Thus, if a cationic polymer is oxidized to smaller cationic subunits, it will still react. However, reactions that render cationic polyelectrolytes neutral or anionic will be reflected as a reduction in polyelectrolyte concentration. In addition, if NOM is present in a cationic form, it can introduce an interference in the test. Depending on the water source pH and alkalinity, NOM tends to be either neutral or anionic, thus reducing the potential for interference.

Example results of measuring the concentration of a polyDADMAC (Catfloc TL) are shown in Figure 3.2. The polyelectrolyte was mixed with Nanopure water at various concentration levels. The samples were measured initially after mixing and then after 18 hours of storage to evaluate the stability of the solutions. The TOC of each solution was also measured. As shown in Figure 3.2, there was minimal change in the recovery of polyelectrolyte from solution due to storage. There was not a significant difference between the measurements of the samples (p-value of 0.9144 at the 95% confidence level). The method appears to be useful for measuring concentrations of polyDADMAC above 0.5 mg/L.

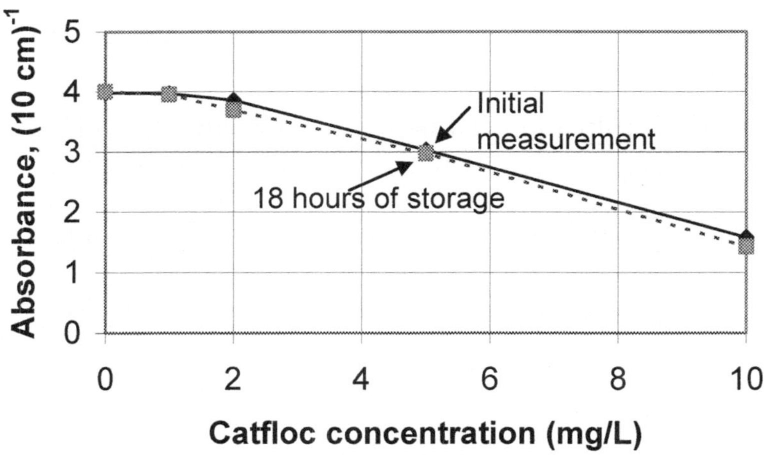

Figure 3-2. Comparison of the use of Ponceau S dye to quantify polyDADMAC solutions before and after 18 hours of storage

The results of the colorimetric method are compared to TOC values in Figure 3.3. As shown, there is a strong correlation between the two tests (R^2 of 0.99). However, for water containing NOM, the TOC test is less useful for quantifying the presence of residual polyelectrolytes.

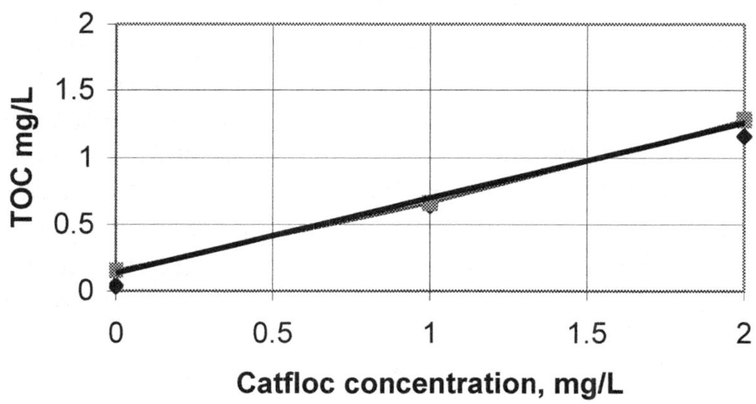

Figure 3-3. Comparison of TOC and colorimetric measurements of polyDADMAC solutions.

Streaming Current Potential. Batch tests for streaming current analysis were conducted by injecting a 50 mL sample into the streaming current detector. Streaming current measurements of cationic polyelectrolyte (Catfloc) are shown in Figure 3.4. As shown, a positive read-out was associated with the cationic polymer and the streaming potential increased with increasing concentration, but the increase was not linear. Concentrations as low as 0.25 mg/L were detected by the streaming current detector. For purposes of comparison, the laboratory tap water (ground water) had a reproducible SCD of -0.05 V.

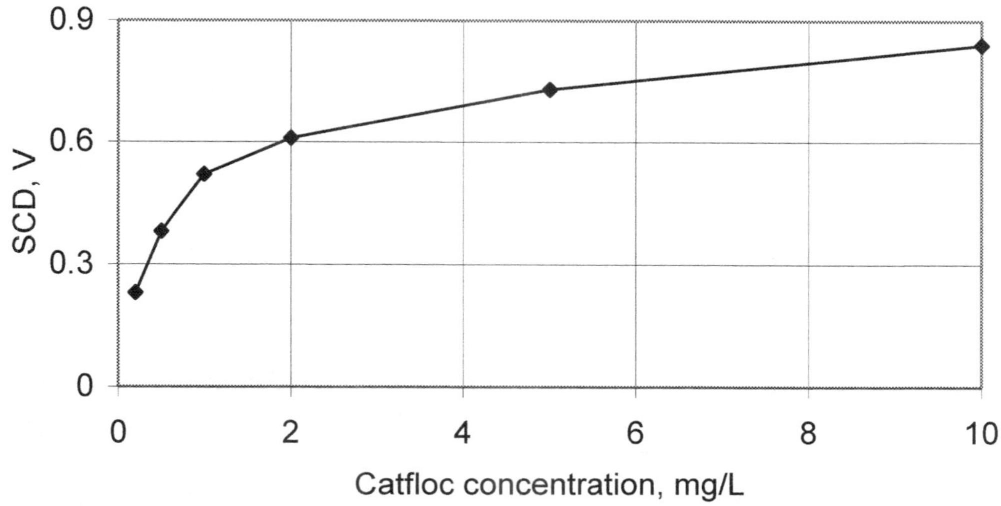

Figure 3-4. Streaming current potential of various concentrations of polyDADMAC polyelectrolyte (Catfloc).

Results of streaming current measurements of anionic polyelectrolyte are shown in Figure 3.5. In this case the signal was negative, corresponding to the anionic nature of the polymer. Similar to measurement of cationic polyelectrolytes, the signal was not linear with concentration. However, Streaming Current measurements can provide a useful screening tool for detecting the presence of unreacted or residual polyelectrolyte and for identifying qualitative changes in the net charge associated with polyelectrolyte addition.

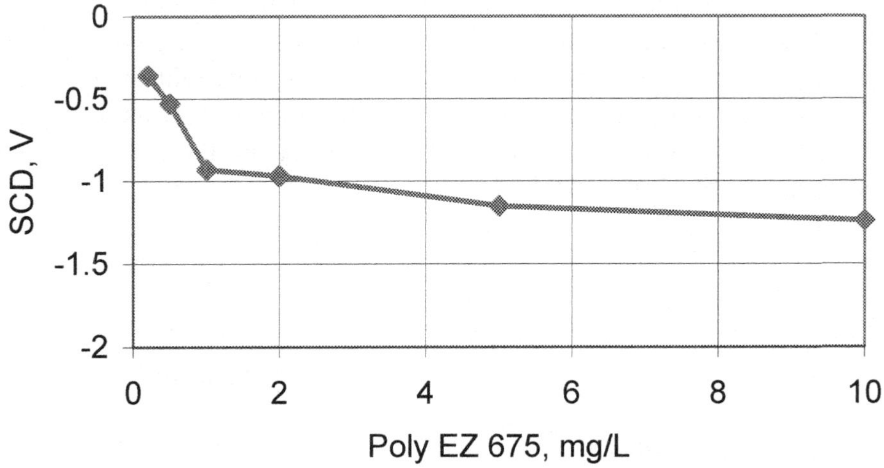

Figure 3-5. Streaming current potential of various concentrations of anionic multi-component polyelectrolyte (Poly EZ 675).

Polyelectrolyte Oxidizability by UV/Persulfate Oxidation. The oxidizability of polyelectrolytes was tested by comparing the extent of oxidation by UV alone and UV coupled with persulfate oxidation. Tests were conducted using a Sievers Total Organic Carbon Analyzer. These tests were used as a screening tool to identify differences among the polymers in terms of oxidizibility and reactivity. A summary of the results of testing 9 different polyelectrolytes is given in Table 3.9. The percent of organic carbon in the polyelectrolytes that was susceptible to UV oxidation varied from 70 percent to 100 percent, depending on the molecular size, charge density, and chemical composition of each polymer.

A comparison of the degree of carbon dioxide production from oxidation using UV alone and in combination with persulfate is shown in Figure 3.6 for two of the compounds screened, a polyDADMAC (Catfloc) and a EPI-DMA (C-572). For the polyDADMAC, UV oxidation was equally effective alone or in conjunction with persulfate. However, for the EPI-DMA, the extent of oxidation was significantly greater when UV was used in the presence of persulfate.

Table 3-9
Comparison of oxidation of organic polyelectrolytes screening test

Trade name	Vendor	Chemical composition	UV oxidizable TOC, mg Carbon/mg polymer	UV/Persulfate oxidizable carbon, mg Carbon/mg polymer	Percent UV oxidizable TOC
SUPERFLOC C-572	Cytec	EPI-DMA	0.348	0.498	70
Percol LT22S	Ciba	Quarternary acrylate salt and acrylamide	0.382	0.495	77
SUPERFLOC N-1986	Cytec	Polyacrylamide	0.457	0.514	89
POL-E-Z 675	Calgon	"Multicomponent liquid"	0.429	0.459	93
Nalclear 8170	Nalco	Acrylamide	0.427	0.443	96
Clarifloc C-3223	Polydyne	Mixture	0.449	0.469	96
CATFLOC TL	Calgon	Poly(dimethyldiallyl-ammonium chloride)	0.449	0.469	96
3895 PWG	Sweetwater	Mixture	0.312	0.320	97
HYPERFLOC AF 300	Polydyne	Polyacrylamide	0.021	0.021	100

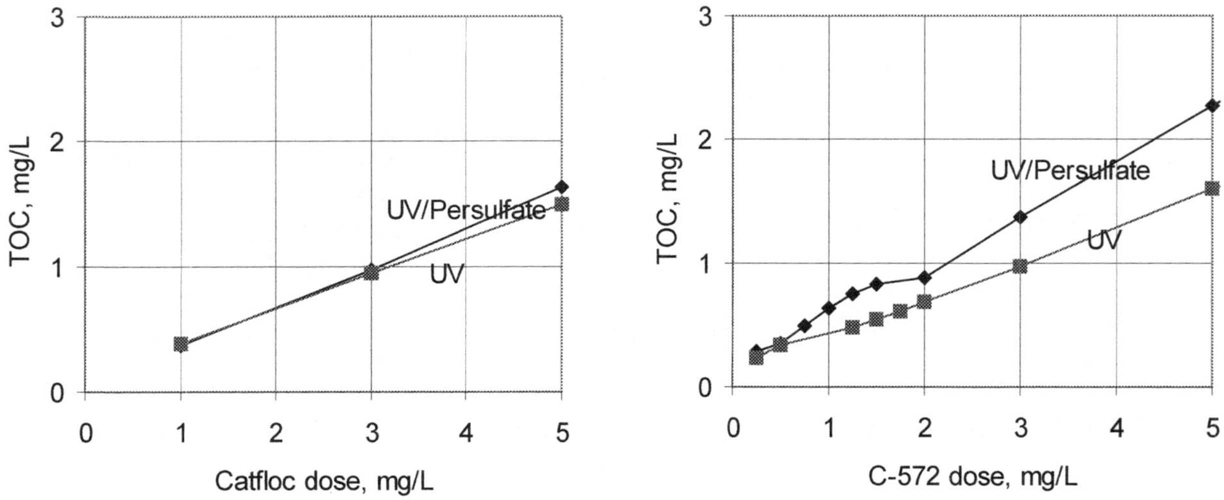

Figure 3-6. Comparison of the effectiveness of UV oxidation alone and in conjunction with persulfate for conversion of polyelectrolyte organic carbon to carbon dioxide.

Evaluation of the stability of UV spectra of polyelectrolytes. Another screening test of the reactivity of polyelectrolytes to UV irradiation was developed using a scanning UV/VIS spectrophotometer. For this test, samples of polyelectrolyte ranging in concentration from 0.1 to 1 mg/L were placed in quartz cells with a path length of 10 cm. Samples were scanned using a Hach DR 4000U spectrophotometer over the wavelength range of 190 to 300 nm. The UV lamp is deuterium and the instrument has a photometric range of 0 to 3 absorbance units and a reproducibility of 0.001 absorbance units. The wavelength accuracy of the instrument is reported to be +/- 1 nm, the resolution is 0.1 nm and the reproducibility is +/- 0.1 nm. The scan speed is 200 nm per minute at 1 nm steps and 20 nm/min at 0.1 nm steps. The spectral bandwith is 4 nm.

The test was conducted by scanning each solution at 1 nm steps, keeping the sample exposed to UV at a fixed wavelength of 254 nm for 10 minutes, zeroing the instrument, and repeating the scan. The total exposure of each sample to UV ranged from 30 to 60 minutes. The UV spectra were evaluated to identify potential changes in UV absorbance as a function of exposure time. The UV spectra for each polyelectrolyte had a similar pattern with the maximum absorbance associated with wavelengths below 200 nm. The magnitude of UV absorbance as a function of concentration varied among the polyelectrolytes tested, depending on the polyelectrolyte composition.

To facilitate comparison of polyelectrolytes, the UV absorbance was normalized to the solution concentration (absorbance divided by concentration L/mg-m). A summary of normalized UV absorbance for several polyelectrolytes is presented in Table 3.10 in order of decreasing UV absorbance. For the polymers tested in this test, the normalized absorbance at 195 nm ranged from 3.5 to 8.3 L/mg-m.

Table 3-10
Normalized UV absorbance of polyelectrolytes at 195 and 200 nm

Trade name	Vendor	Chemical compostion	UV absorbance @ 195 nm L/mg-m	UV absorbance @ 200 nm L/mg-m
CATFLOC TL	Calgon	Poly(dimethyldiallyl-ammonium chloride)	8.3	6.1
SUPERFLOC N-1986	Cytec	Polyacrylamide	7.7	4.7
Clarifloc C-3223	Polydyne	Mixture	7.6	4.8
SUPERFLOC C-572	Cytec	EPI-DMA	5.8	4.2
HYPERFLOC AF 300	Polydyne	Polyacrylamide	5.4	3.7
Nalclear 8170	Nalco	Polyacrylamide	4.9	2.7
Percol LT22S	Ciba	Quarternary acrylate salt and acrylamide	3.6	2.2

The stability of the UV spectra following continuous exposure to UV in the spectrophotometer was evaluated as a screening tool. For the majority of the polyelectrolytes tested, the spectra were relatively consistent for up to a 60 minute exposure period. However, for two of the polyelectrolytes, there appeared to be a decrease in UV absorbance at the lower wavelengths after about 10 minutes of UV exposure. UV spectra for a polyDADMAC (Catfloc) and an EPI-DMA (C-572) are shown in Figure 3.7. Exposure beyond 20 minutes did not result in additional changes in the spectra.

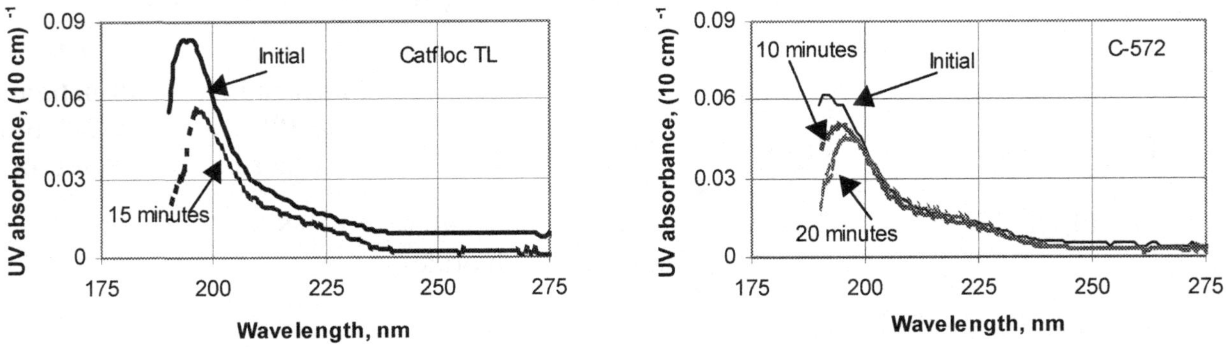

Figure 3-7. Comparison of the UV-spectra for two polyelectrolytes, a PolyDADMAC (Catfloc TL) and an EPI-DMA (C-572). Each graph represents 0.1 mg/L of polyelectrolyte measured using a 10 cm quartz cell under continuous exposure to UV light.

Prechlorination of polyelectrolytes. The impact of prechlorination on oxidation of polyelectrolytes was investigated as an additional screening test for polyelectrolyte reactivity. For this test, a fixed concentration of each polymer (5 mg/L) was exposed to chlorine at dosages ranging from 0 to 10 mg/L. The chlorine exposure time was 24 hours. The samples were then analyzed in the TOC analyzer with UV/persulfate oxidation. The carbon dioxide formed from each polymer was measured as TOC. A comparison of the response of four polyelectrolytes is shown in Figure 3.8. For the majority of the compounds tested, chlorine had minimal effect on the carbon dioxide produced from UV/persulfate oxidation. However, in the case of the EPI-DMA (C-572), prechlorination resulted in an apparent decrease in the sample TOC. It is unknown if the decrease in TOC is due to formation of compounds that are resistant to UV/persulfate oxidation or if chlorine reactions resulted in conversion of a portion of the polyelectrolyte to carbon dioxide prior to UV/persulfate oxidation.

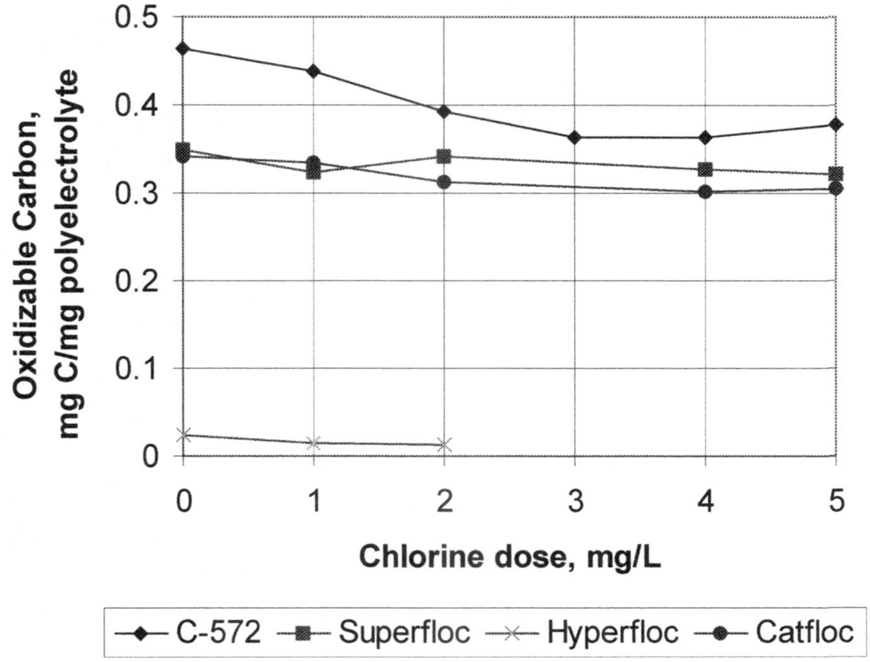

Figure 3-8. Impact of chlorine dose on the oxidizable carbon content of four polyelctrolytes: an EPI-DMA (C-572), a polyDADMAC (Catfloc) and two polyacrylamides (Superfloc and Hyperfloc).

Capillary Electrophoresis. Capillary electrophoresis was used to quantify residual polyelectrolytes. All tests were conducted using a Beckman PACE 5000 capillary electrophoresis system. The capillary used in these tests is a 57 cm uncoated fused-silica CE column (75µm I.D., 300 µm O.D., 50 cm length to the detector). A summary of the conditions tested is given in Table 3.11. The major variables tested were the buffer composition. Over 800 trial tests were conducted to optimize the conditions and ensure that reproducible results could be obtained. The test protocol that was adopted is shown in Table 3.12.

Table 3-11
Summary of variables tested for capillary electrophoresis separation of polyelectrolytes

Parameter	Range	Rationale
Wavelength (nm)	200, 214	Optimize sensitivity, detection limits, minimize interferences from buffers
Separation voltage (kV)	-30, +10, +25, +30	Intensity of voltage is related to migration time; polarity of voltage is related to net charge on the molecule
Separation time, min	10, 15.	Separation time increases with decreased voltage
Temperature, ^{o}C	20, 25 , 30 , 35 , 40 .	The rate of separation is influenced by temperature. If temperatures are too high, there may not be adequate separation, whereas temperatures that are too low may result in poorly defined electrophoretic peaks.
Injection time @ 0.5 psi, sec	5, 7, 10	Injection time is proportional to the volume of sample injected.
Buffers	Phosphate; borate	These buffers are widely used in capillary electrophoresis. The pK_a and pH of the buffer influences the net charge on the polylectrolyte as it passes through the capillary.

Table 3-12
Summary of Test Conditions for Capillary Electrophoresis of Organic Polyelectrolytes

Parameter	Values
Buffer	Boric Acid: pH 9; 10- 20 mM
Voltage	25K
Injection	High pressure, 7 second injection
Duration	10 minutes
Separation conditions	Voltage applied, forward
Protocol	0.1% NaOH: 1 minute Water: 2 minutes Buffer: 2 minutes Injection: 10 sec Buffer: 10 minutes Water: 2 minutes

Test results were evaluated in terms of reproducibility, the consistency of polyelectrolyte migration time and the relationship between peak area and polyelectrolyte concentration. The three different polyacrylamide reference standards that were used to validate the method are listed in Table 3.13. One standard was an acrylamide monomer, the other standards were purified polyacrylamides. These standards provided a basis for comparing the commercial preparations used by water treatment facilities.

Table 3-13
Chemicals used for standardization of Capillary Electrophoresis Testing

Chemical	Composition	Molecular weight, amu	Source
Poly(acrylamide-co-acrylic acid)	1.5% Acrylic acid	5,000,000	Aldrich Chemical
Poly(acrylamide-co-acrylic acid)	partial sodium salt		Adrich Chemical
Acrylamide	C_3H_5NO	71.08	Sigma Chemical

The separation conditions for each formulation of polyelectrolyte vary depending on the composition, molecular size and net charge. A systematic approach was used to optimize the separation of each polyelectrolyte with the major variables being the concentration of separation buffer and the separation temperature and voltage. Because the capillary electrophoresis separation is complete within a few minutes, it was possible to develop and verify the test protocol for each chemical within a few hours.

The primary focus of work in this project was on analysis of acylamide based polyelectrolytes. A list of the formulations tested in given in Table 3.14. For each formulation, a capillary electrophoresis method was developed and validated. Then disinfection byproduct formation potential tests were conducted and the residual solution was characterized using capillary electrophoresis. An example electropherogram for Cytec A-100 is shown in Figure 3.9. The detection limits for this method were about 2.5 mg/L.

Table 3-14
Polyelctrolytes tested for capillary electrophoresis verification

Polymer trade name	Vendor	Components (from Label)	Function	MSDS #
Superfloc N-300	Cytec	Polyacrylamide, Acrylamide homopolymer, Ammonium chloride	Flocculant	9156
Superfloc A-130	Cytec	NJ Trade Secret Registry #00901401000-5043 P, 00901401000-5000 P	Flocculant	6913
Superfloc A-100	Cytec	Anionic Polyacrylamide, sodium acrylate	Flocculant	1782
Superfloc AF 126	Cytec	Ammonium polyacrylate in water-in-oil emulsion, ammonia, ammonium acetate	Filter Aid	9770
N8170	Nalco	Acrylamide	Flocculant	

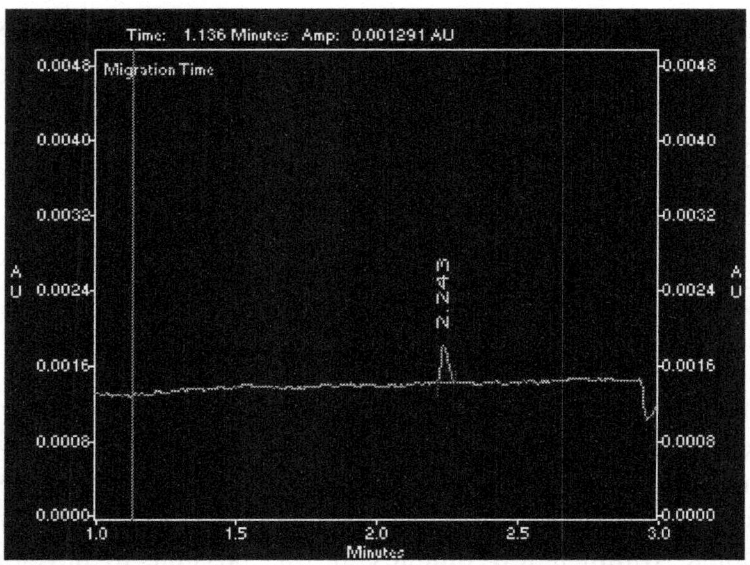

Figure 3-9. Example output for separation of 5 mg/L of Cytec A-100 using capillary electrophoresis with 20 mM Boric Acid buffer at pH 9.

Size Exclusion Chromatography. The use of size exclusion chromatography (SEC) was tested in parallel with capillary electrophoresis to provide an independent measure of polyelectrolyte characteristics. Tests were conducted using the method reported by Soponkanaporn and Gehr (1989). The tests were conducted on polyacrylamide polymers. Example results of testing a anionic polyacrylamide (A-130) and a nonionic polyacrylamide (N-300) are shown in Figure 3.10. Without oxidation, each polymer had a single peak at a retention time of about 8.3 minutes. Following chlorine oxidation, two peaks were present, suggesting that a lower molecular weight subunit had formed from the oxidation reaction. Permangante alone did not appear to influence the separation properties of polyacrylamides. In addition, it was not possible to differentiate between multiple polymers in a single water. Based on testing conducted in this project, size exclusion chromatography was less sensitive than capillary electrophoresis and this method was not used for other components of the project.

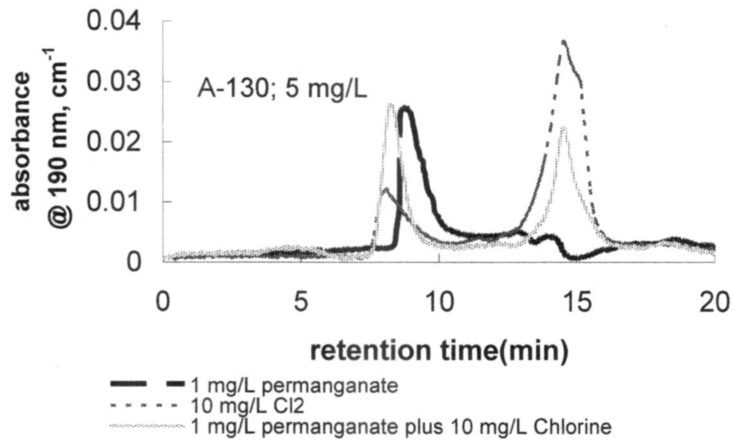

a) Superfloc A-130 pretreated with permanganate, chlorine, or permanganate and chlorine.

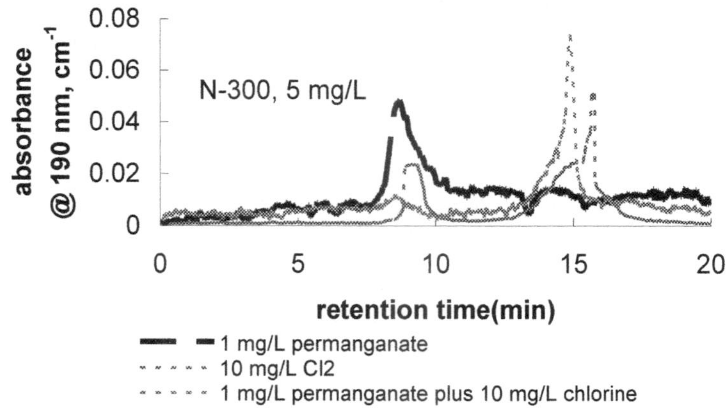

b) Superfloc N-300 pretreated with permanganate, chlorine, or permanganate and chlorine.

Figure 3-10. Comparison of size exclusion chromatography results for oxidation of two polyacrylamide polyelectrolytes treated with permanganate, chlorine, or permanganate and chlorine a) Superfloc A-130; b) Superfloc N-300.

Bench-Scale Chemical Interactions

An overview of potential chemical interactions is given in Table 3.15. Based on chemical use practices in the water industry and the literature review (Chapter 1), the predominant focus of this project was on potential reactions of polyelectrolytes with multiple oxidants.

Table 3-15
Overview of Chemical Interactions

Polyelectrolyte	Oxidant	Coagulant	Disinfectant	Other
Cationic	Permanganate	Alum	Chlorine	Sorbents: PAC
Anionic	Ozone	Ferric chloride	Ozone	Corrosion inhibitors
Nonionic	Hydrogen peroxide	Polyaluminum chloride	UV	Fluoridation agents
	Chlorine	Other inorganic coagulants	Combined chlorines	Lime
	UV		Chlorine dioxide	Sodium hydroxide
	Mixed oxidants		Mixed oxidants	

Performance Testing

Polyelectrolyte performance was evaluated in the absence and presence of other water treatment chemicals. Each polyelectrolyte was tested directly and following UV exposure. The influence of oxidants on polyelectrolyte performance was tested by adding a fixed amount of either chlorine (5 mg/L) or permanganate (0.8 mg/L) to each reactor. The combined effects of chlorine and UV were also evaluated. The test conditions are summarized in Table 3.16.

Table 3-16
Summary of test conditions used for polyelectrolyte performance testing

Parameter	Description
Polyelectrolytes	Catfloc TL, Superfloc N-300
Dosage range	0 to 1 mg/L of each polyelectrolyte
Oxidant chemicals	Potassium permanganate, 0.8 mg/L; Chlorine, 5 mg/L
UV irradiation	Irradiate polyelectrolyte prior to addition to each reactor by exposing solution to UV light at 254 nm for 30 minutes.

A standardized procedure was developed to produce a particle suspension for all laboratory screening tests conducted in this project. Kaolinite (40 g) was mixed with 6% sodium hydroxide. The mixture was heated, cooled, washed with 1 M sodium chloride, and settled. The suspension was washed with Nanopure water until the conductivity was below 1100 μS/cm. Then the suspension was allowed to stabilize for 24 hours. Aliquots of the suspension were diluted in accordance with the objectives of the laboratory tests, diluted samples are amended with minerals (alkalinity, hardness), the pH was adjusted and then the suspension was used in polyelectrolyte performance tests.

Polyelectrolyte mixing procedures were adapted from Dental (1995). The chemicals were mixed in plastic Nalgene containers for one hour at a G value of about 500 sec $^{-1}$. The solutions were then aged overnight and mixed for another hour at a G value of about 500 sec^{-1}. Polyelectrolyte concentrations were measured using TOC and the concentrations of cationic polymers were confirmed using the Ponceau S dye method.

The polyelectrolyte performance tests were developed to relate performance indicators such as turbidity removal, particle size distributions, and streaming current analysis to the dosage. Several rounds of jar testing were conducted to evaluate the effectiveness of using particle size and streaming current data in conjunction with conventional end-points (turbidity, color) to identify overdosages of polyelectrolyte. The test matrix consisted of 200 mL of the Kaolinite suspension diluted to one liter and amended with sodium bicarbonate. PolyDADMAC polymer (CATFLOC) was added to each reactor. Reactors were rapid mixed at a G value of 300 sec^{-1}, and then mixed using a decayed gradient (sequential G-values of 55, 33, and 14 sec^{-1}) for a total flocculation time of 30 minutes. After 30 minutes of sedimentation, the supernatants were evaluated for turbidity, particle size distributions, residual polyelectrolyte, and streaming current potential.

The use of particle size distributions was evaluated as a measure of polyelectrolyte performance. The median particle size and the relative number of particles remaining suspended in solution (after sedimentation) are shown in Figure 3.11 for two tests using the Kaolinite solution reacted with polyDADMAC polymer. While there was some variability in the number of particles remaining at each dose, the trends for the two sets of samples were similar. At a polymer dose of about 2 mg/L the median particle size was reduced to below 5 μm. However, the most significant reduction in the number of particles occurred at a dose of about 1 mg/L.

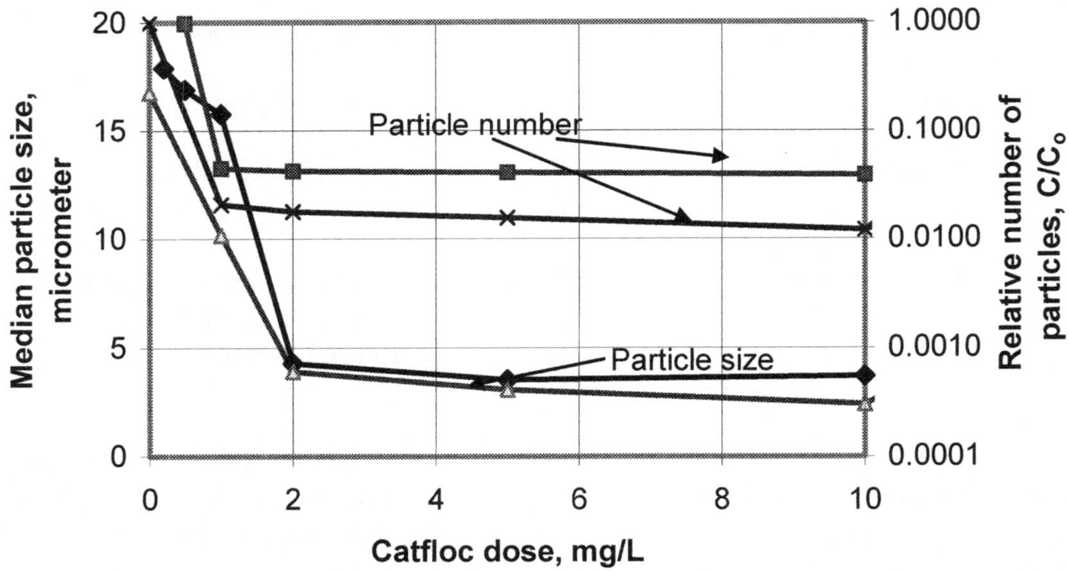

Figure 3-11. Influence of polyelectrolyte dose on median particle size and relative number of particles remaining suspended in solution after coagulation-flocculation-sedimentation in a jar test of a kaolinite suspension with an initial turbidity ranging from 320 to 365 NTU (data represent the results from two sets of tests).

Complementary data on turbidity, streaming current potential, and residual polyelectrolyte concentration are provided in Figure 3.12. As shown, the optimum polyelectrolyte dose based on turbidity reduction was about 6 mg/L with sharp increases in turbidity at dosages above 10 mg/L. Residual polyelectrolyte was detected in the solution at dosages over 2 mg/L. It is interesting to note that the concentration of residual polyelectrolyte dissolved in the samples appears to be constant at dosage values over 5 mg/L. It is likely that some of the excess polyelectrolyte combined with the kaolinite and therefore was not measured in the solution.

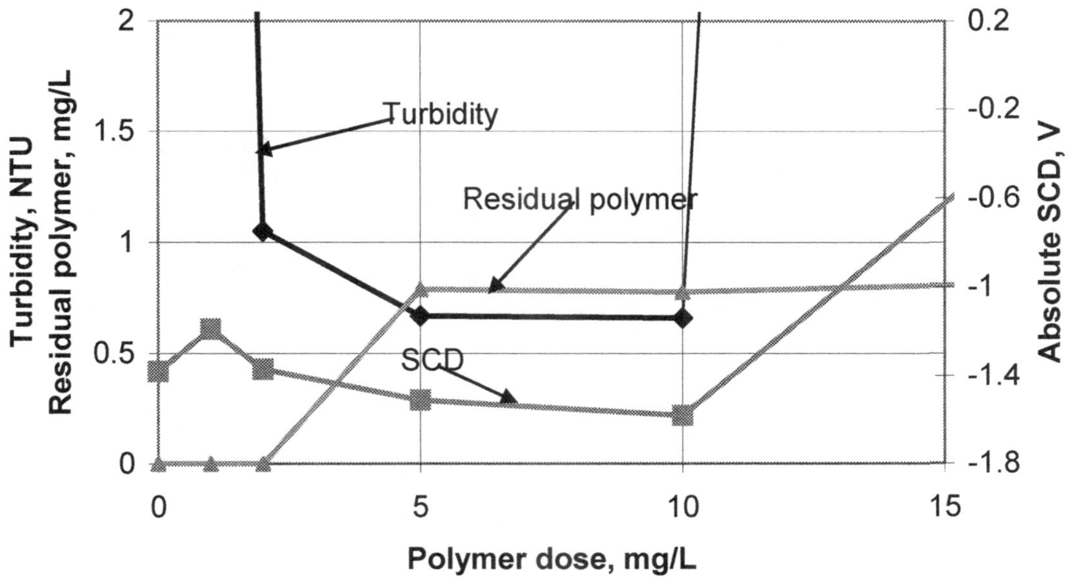

Figure 3-12. Influence of polyelectrolyte dose (Catfloc) on residual turbidity, residual polyelectrolyte concentration, and streaming current after coagulation-flocculation-sedimentation in a jar test on a kaolinite suspension with an initial turbidity of 363 NTU.

The streaming current signal remained negative over the entire dosage range, suggesting that the net negative charge on the particles had a more dominant control over the solution properties than the positive charges introduced by the addition of polyelectrolyte. An increase in streaming current detection did correspond to an increase in turbidity but did not correlate to the presence or absence of residual polyelectrolyte. Based on these findings, it is obvious that no single measure of performance (particle size, particle number, turbidity, streaming current, etc.) can provide a complete description of the process.

For each test, the polylelectrolyte dose influenced the finished water turbidity, particle size distribution, and settling velocities. The impacts of chlorine, permanganate, and UV exposure can be seen most clearly by comparing the finished water turbidity and time-to-filter for each polyelectrolytes. Results for testing of a polyacryamide flocculant (N-300) are shown in Figure 3.13 for finished water turbidity. When residual turbidity is used as a performance measure, all treatments yielded about the same optimum dose of flocculant, however, the lowest residual turbidity was associated with using the polyelectrolyte in the absence of oxidant chemicals. Exposure of the polyelectrolyte to UV irradiation did not result in a significant difference in the residual turbidity as compared to no UV exposure. However, chlorine, permanganate and chlorine combined with UV exposure all resulted in decreased performance as measured by residual turbidity. Similar results were obtained for other polyelectrolytes, however the magnitude of difference among different treatments was dependent on the polyelectrolyte.

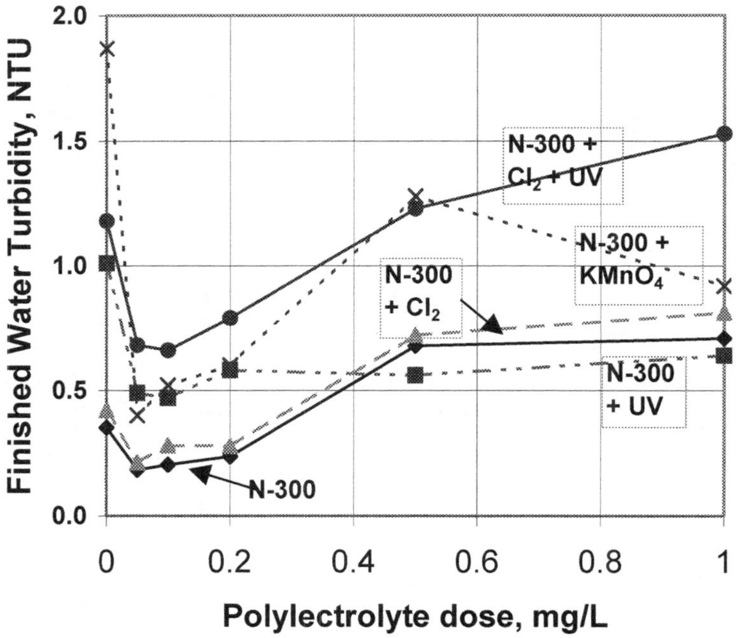

Figure 3-13. Comparison of finished water turbidity of Kaolinite suspensions treated with Cytec flocculant N-300 (polyacrylamide) under different conditions following coagulation/flocculation and sedimentation. The chlorine dose used was 5 mg/L and the permanganate dose was 0.8 mg/L. The UV exposure was conducted using a UV chamber at a wavelength of 254 nm.

A different perspective on polyelectrolyte performance is observed by comparing the time-to-filter associated with each treatment. Results for treating the Kaolinite suspensions with Cytec N-300 flocculant are shown in Figure 3.14. Exposure to UV and/or chlorine or permanganate resulted in an improvement in the "filterability" of the water as measured by the time-to-filter. The poorest performance was associated with the use of the polyelectrolyte in the absence of UV or oxidants. In this case the samples with the lowest turbidity (see Figure 3.13) required the longest time-to-filter (about a two-fold difference).

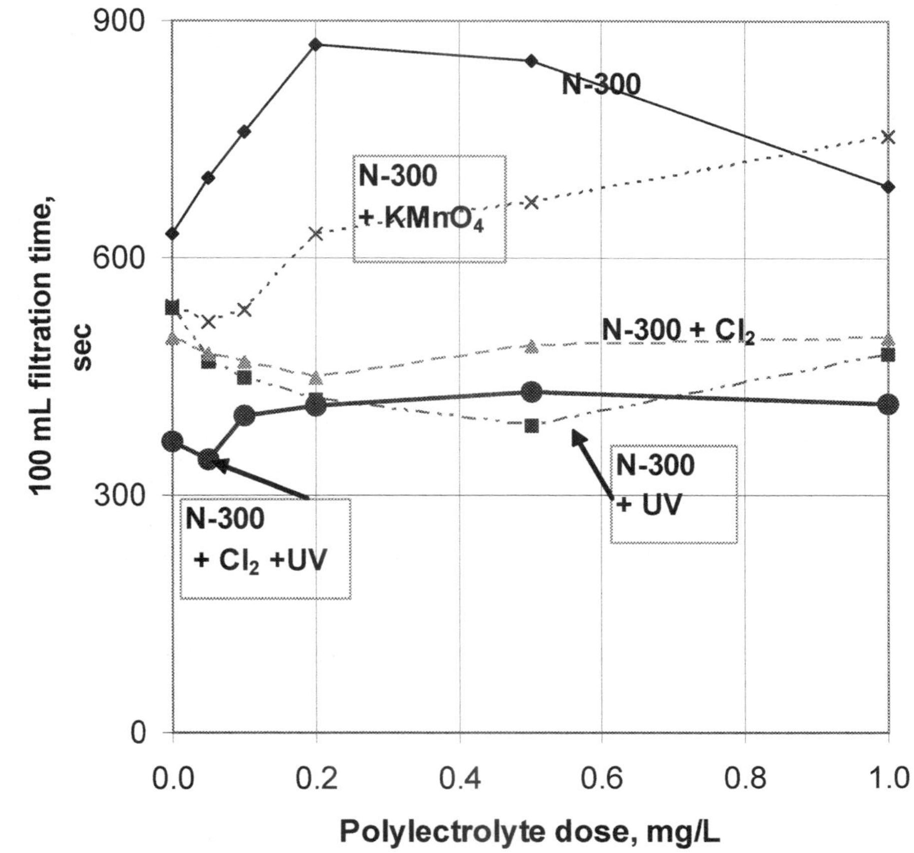

Figure 3-14. Comparison of the time-to-filter 100 mL of Kaolinite suspension treated with Cytec flocculant N-300 (polyacrylamide) under different conditions following coagulation/flocculation and sedimentation. The chlorine dose used was 5 mg/L and the permanganate dose was 0.8 mg/L. The UV exposure was conducted using a UV chamber at a wavelength of 254 nm.

Testing of Inorganic Coagulant Interactions

Inorganic coagulant interactions were tested using the model particle suspension and a surface water. The model suspension was tested using a jar test apparatus to evaluate the impacts of ferric sulfate and polyelectrolyte on particle size distributions. Coagulant chemicals and polyelectrolytes were added to each reactor under conditions of rapid mixing. Then the mixing speed was reduced to simulate a three stage decayed gradient flocculation period. After flocculation, a 60 minute quiescent period was used to simulate sedimentation. Water quality analyses were conducted prior to the test and at intervals during the test. Particle size distributions were measured using a Spectrex laser light scattering instrument (PC-2000). A comparison of the particle size distribution of the suspension after coagulation, flocculation, and sedimentation using ferric sulfate, or ferric sulfate plus polyelectrolyte is shown in Figure 3.15. For these tests, there was no significant difference in the finished water particle size distribution.

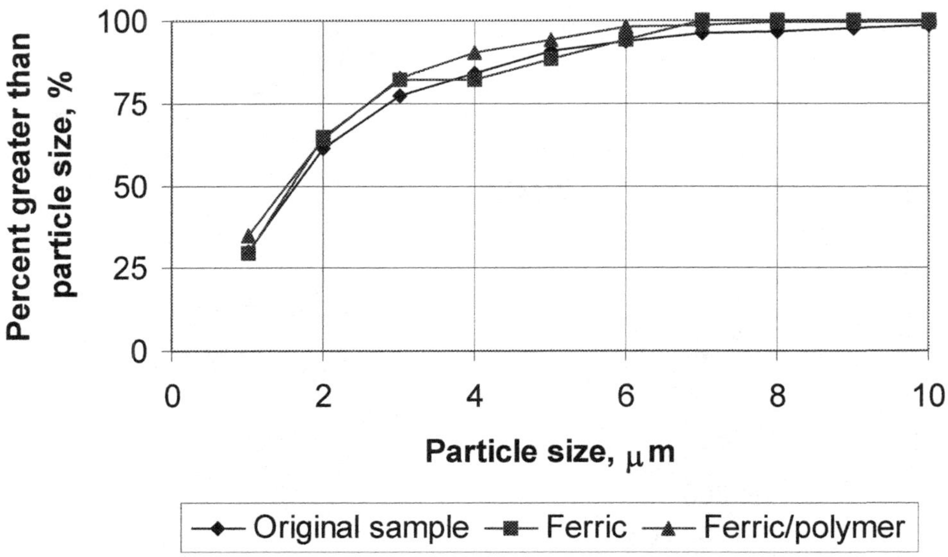

Figure 3-15. Comparison of the particle size distribution of a model kaolinite suspension following coagulation with ferric sulfate or ferric sulfate and polymer.

A surface water was used to provide a natural water source for assessing the reproducibility of the test methods and to identify potential interferences due to NOM and other constituents. The water was obtained from the Hillsborough River, the water supply for the City of Tampa, Florida. The characteristics of the water sample are shown in Table 3.17. The TOC was not measured in these samples, but it typically ranges from 3.5 to 8 mg/L during this season.

Table 3-17
Water quality parameters of Hillsborough River water [1]

Parameter	Value
Alkalinity	144 mg/L as $CaCO_3$
Color	60 color units
Conductivity	404 μS/cm
Dissolved oxygen	8.2 mg/L as O_2
pH	8.6
SCD	-0.12 V
Turbidity	4.55 NTU

[1]Testing conducted in April 2000

An initial round of jar tests were conducted using ferric sulfate to determine the optimum dose for removal of turbidity and color from this water. Based on the preliminary tests, a dose of 100 mg/L $Fe_2(SO4)_3$ was used in the tests with polyelectrolyte. A set of six reactors was used and ferric sulfate was added concurrently with various dosages of polyelectrolyte (CATFLOC). Samples were mixed at a G value of 300 sec^{-1} for one minute, subjected to a decayed gradient flocculation period of 30 minutes, and settled for 60 minutes. The residual turbidity and SCD were measured. Results are shown in Figure 3.16. The "zero" dose refers to the use of ferric sulfate without polyelectrolyte. In the case of natural water, other constituents may contribute to the net electrical potential, making the streaming current signal difficult to interpret.

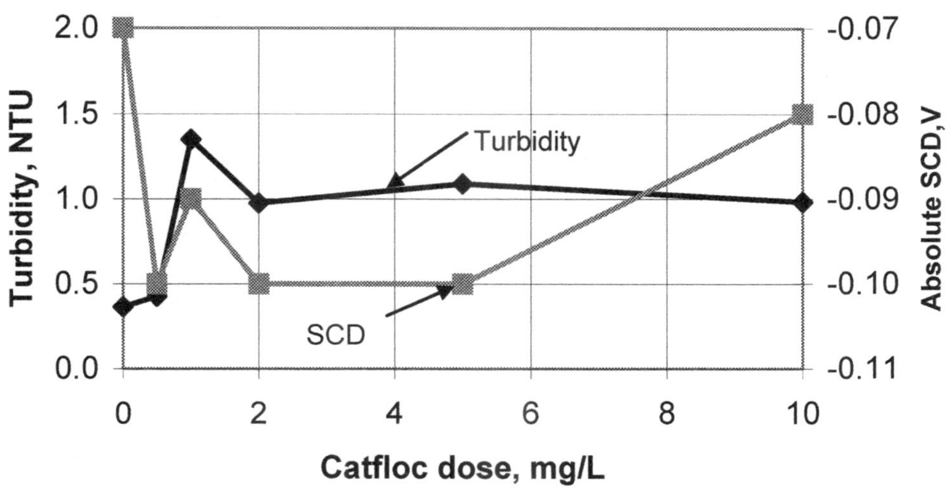

Figure 3-16. Influence of polyelectrolyte dose (Catfloc) on residual turbidity and streaming current after coagulation-flocculation-sedimentation in a jar test on Hillsborough River water. Polyelectrolyte was added concurrently with 100 mg/L of ferric sulfate to each reactor.

Testing of oxidant interactions in the presence and absence of ultraviolet light

The primary focus of the testing was on UV interactions with polylectrolytes in the presence of chlorine. Extensive testing was conducted using two polyelectrolytes Hyperfloc 304 and Catfloc TL. For each polyelectrolyte, stock solutions were prepared at a concentration of 1 mg/mL. Working solutions were prepared by diluting the stock solutions with Nanopure water at a mixing speed of 300 rpm. The dilution procedure yielded concentrations ranging from 0.2 to 5 mg/L. Chlorine was added to the diluted polymer prior to exposure to either UV irradiation or sunlight. The samples were irradiated either by direct sunlight or using a Millipore UV disinfection chamber. The light intensity was measured using a 1400A Radiometer/Photometer instrument. For tests involving UV irradiation, the polyelectrolyte was irradiated prior to injection into the reactors. A syringe was used to add the polyelectrolyte to each reactor. A matrix of the test conditions is given in Table 3.18.

Table 3-18
Summary of UV irradiation test conditions

Polyelectrolyte mg/L	UV irradiation	Chlorine	UV irradiation + chlorine	Sunlight	Sunlight + chlorine
0	-	5 mg/L	-	-	-
0	30 min	-	-	-	-
0	-	-	30 min, 5 mg/L	-	-
0	-	-	-	-	30 min, 5 mg/L
2.0	-	5 mg/L	-	-	-
2.0	30 min	-	-	-	-
2.0	-	-	-	30 min	
2.0	-	-	30 min, 5 mg/L	-	-
2.0	-	-	-	-	30 min, 5 mg/L

The influence of ultraviolet light on polyelectrolytes was tested in several ways. A comparison of the chlorine demand due to UV exposure of Nanopure water containing 2 mg/L of polyelectrolytes is shown in Figure 3.17. As shown, the polyacrylamide (Hyperfloc) had a slightly higher chlorine demand than the polyDADMAC (CatflocTL). Direct exposure to UV irradiation imposed a slightly higher degree of chlorine demand than did sunlight.

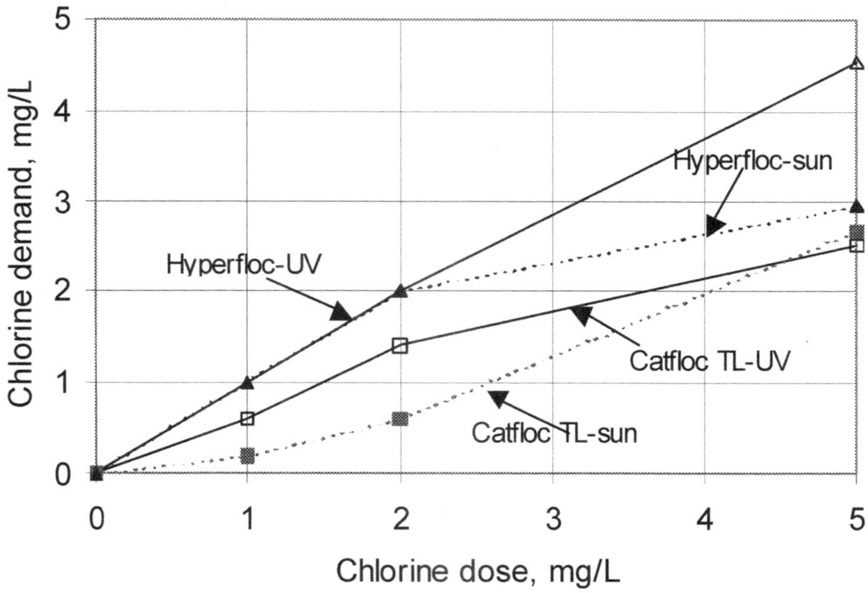

Figure 3-17. Comparison of chlorine demand due to exposure of 2 mg/L of polyelectrolytes (Catfloc TL and Hyperfloc 304) to 30 minutes of either sunlight or UV. The UV dose from sunlight was about 9 mJ/cm^2.

Residual polyelectrolyte concentrations following exposure to sunlight or UV light in the presence or absence of chlorine in Nanopure water are summarized in Table 3.19. As shown, there is significant reduction in the concentration of residual polyelectrolyte following exposure to sunlight. The combination of sunlight (or UV) and chlorine may result in an advanced oxidation process that appears to reduce the residual polyelectrolyte concentration.

Table 3-19
Comparison of Residual Polyelectrolyte Concentrations due to exposure to sunlight, UV irradiation and/or chlorine

Sample	Catfloc TL, mg/L	Hyperfloc, mg/L
Initial concentration, mg/L	2.4	2.1
Polymer + UV (30 min)	0.5	ND[a]
Polymer + 5.0 mg/L Cl_2 + 10% excess Na_2SO_3	ND[a]	2.1
Polymer + 5.0 mg/L Cl_2 + UV (30 min) + 20% excess Na_2SO_3	ND[a]	ND[a]
Polymer + 5.0mg/L Na_2SO_3 + UV (30 min)	0.6	ND[a]
Polymer + sunlight (30 min)	0.04	ND[a]
Polymer + 5.0mg/L Na_2SO_3 + sunlight (30 min)	ND[a]	ND[a]
Polymer + 5.0 mg/L Cl_2 + sunlight (30 min) + 20% excess Na_2SO_3	ND[a]	ND[a]

[a] ND represents Not Detected by the Ponceau dye method

Impacts of sunlight on polyelectrolyte properties

To determine the impacts of sunlight on polyelectrolyte properties, controlled solutions of polymer were mixed with Nanopure water using glass beakers that had been acid washed and oven dried. The nominal concentration of polymer in each sample was 5 mg/L. Alkalinity was added to selected reactors to provide 100 mg/L as $CaCO_3$. Sodium hypochlorite was added to selected reactors at a dose of 2 mg/L as Cl_2. Initially, the pH, alkalinity, chlorine, and polymer concentration of each sample was recorded. Samples were placed in direct sunlight on the roof of the Kopp Engineering building for 30 minutes under maximum ultraviolet exposure (Noon to 2:30 pm, clear sky, > 80° F). The depth of sample in each reactor was 6 cm. Following exposure to sunlight, samples were covered and water quality analyses were conducted. Results for tests conducted on a polyDADMAC are summarized in Table 3.20.

Table 3-20
Results of sunlight irradiation of PolyDADMAC.

Reactor characteristics			pH		Chlorine, mg/L as Cl_2				PolyDADMAC conc., mg/L	
Initial alkalinity, mg/L as $CaCO_3$	Cl_2 dose, mg/L	Sunlight Exposure time, min	Initial	Final	Initial Free	Initial Total	Final Free	Final Total	Initial	Final
Chlorine Reactions (no polyelectrolyte)										
0	2	30	5.9	5.1	2.0	2.0	1.5	1.5		
100	2	30	7.4	7.8	1.7	1.7	1.2	1.2		
Polyelectrolyte Reactions (no chlorine)										
0	0	0	5.7	5.7					4.63	4.63
0	0	30	5.7	4.9					4.63	4.48
Polyelectrolyte and chlorine reactions										
0	2	0	5.7	5.7	0.95	1.5	0.95	1.5	4.63	3.72
100	2	0	7.8	7.8	0.8	1.4	0.8	1.4	4.63	2.35
0	2	30	5.8	4.9	0.95	1.5	ND	ND	4.63	2.82
100	2	30	7.8	6.9	0.8	1.4	ND	ND	4.63	ND

This set of tests provided valuable insight into the chemical interactions that occur in the presence of sunlight (such as during water treatment). First, it is interesting to look at the chlorine distribution. As would be expected, some decay in chlorine occurred in all chlorinated samples due to photochemical reactions (even in the absence of polyelectrolyte). About 0.5 mg/L as Cl_2 was consumed during the 30 minutes of sunlight exposure. A slight reduction in polyelectrolyte concentration occurred following 30 minutes of sunlight exposure. However, the most dramatic reductions in polyelectrolyte concentration occurred in the chlorinated polyelectrolyte samples. Even in the absence of sunlight, significant changes occurred in the polyelectrolyte concentrations, with a higher degree of reaction observed in the samples amended with alkalinity (higher pH). It is also interesting to note that over 35% of the chlorine formed combined chlorine (Total chlorine minus Free chlorine) immediately after chlorine addition to the polyelectrolyte samples.

The coupled effect of chlorine and sunlight can be seen in the last two entries of the Table. All of the chlorine was consumed within the 30 minute reaction time. In the acidic samples (no supplemental alkalinity), the polyelectrolyte concentration decreased to about half of its original value. In the neutral samples, the final polyelectrolyte concentration was below detection limits (0.5 mg/L).

Disinfection By-Product Testing

Polyelectrolytes can play a paradoxical role in the context of disinfection byproduct control. Depending on the form of DBP precursors in water (molecular size, hydrophobicity, etc.) polyelectrolytes can be effective in coagulation of TOC. However, polyelectrolytes also contribute TOC to water and can provide precursor material for formation of DBPs depending on operating conditions. Tests were conducted to evaluate the degree to which cationic polyDADMAC (Catfloc T) can form DBPs due to reactions with permanganate. Water samples were reacted with polymer and permanganate using standardized jar testing procedures. Following a reaction period, each sample was chlorinated and incubated in the dark for 7 days. For samples containing polymer alone, no DBP formation was observed in this test. A comparison of DBP formation for two levels of polymer reacted with permanganate is shown in Table 3.21. These tests provide evidence that some type of oxidation reaction produced constituents that are susceptible to reaction with chlorine.

Table 3-21
7-day DBP formation potential due to polymer and permanganate addition

Disinfection by-product	1 mg/L Polymer + 1 mg/L $KmnO_4$	5 mg/L Polymer + 1 mg/L $KmnO_4$
Total THMs, µg/L	----	6.8
Dichloroacetic acid, µg/L	12.2	29.9
Trichloroacetic acid, µg/L	10.4	35.8
Dichloroacetonitrile, µg/L	0.6	
Chloropicrin, µg/L	---	1.2
1,1,1-trichloropropanone-2	1.6	0.7
1,1-dichloropropanone-2		0.9

In the context of drinking water treatment, it is possible that residual oxidant and polyelectrolyte chemicals may be present in the water when disinfectant chemicals are added. It is important to quantify the conditions under which DBP formation can occur. It is also important to evaluate the extent to which residual polyelectrolyte contributes to overall DBP formation. For the DBP results shown in Table 3.21, formation of HAAs and HANs was more significant the THM formation. However, other polyelectrolytes and other treatment conditions may result in different end points.

The formation of THMs and HAAs was evaluated for three polyelectrolytes: Cytec A-100, Cytec A-133, and Cytec N-300. The polyelectrolyte solutions were prepared using Nanopure water and the pH was adjusted to 7± 0.2, using 0.5 Nsodium hydroxide A summary of pretreatment conditions applied to each polyelctrolyte is given in Table 3.22. Three parallel samples were used for each test.

Table 3-22

Pretreatment of polyelectrolytes prior to disinfection byproduct formation potential testing

Polylelectrolyte concentration	Chlorine dose	Permanganate dose	Sunlight exposure
0, 5 mg/L	0	0	0
0, 5 mg/L	5 mg/L	0	0
0, 5 mg/L	5 mg/L	0	1 hour
0, 5 mg/L	0	0.8 mg/L	0
0, 5 mg/L	0	0.8 mg/L	1 hour

Following the pretreatment step defined in Table 3.22 the disinfection byproduct formation potential test was conducted according to Standard Methods. Chlorine was added to each sample to provide an initial concentration of 10 mg/L. Samples were incubated in the dark at 20 ° C for 7 days. The final chlorine concentration and pH were recorded for each sample. In the vials for DBP analysis, the residual chlorine was quenched using sodium thiosulfate and the samples were transported to the City of Tampa Water Department laboratory for analysis of trihalomethanes and haloacetic acids. As a component of the DBP testing, the City of Tampa Water Department Laboratory ran replicate analyses, quality assurance checks, fortified blanks, and followed rigorous protocols for quality control.

The test results are summarized in Figure 3.18 for Superfloc A-100, Figure 3.19 for Superfloc A-133, and in Figure 3.20 for Superfloc N-100. In all cases, exposure to sunlight and oxidants impacted the net formation of trihalomethanes and haloacetic acids, however the degree to which the disinfection byproduct formation potential was altered was specific to each polyelectrolyte. For Superfloc A-100 and N-300, increased levels of disinfection byproducts were associated with exposure to oxidants and/or sunlight, whereas the highest levels of disinfection byproduct formation for Superfloc A-130 were associated with the polymer alone. It appears that exposure to oxidants and/or sunlight reduced the formation of trihalomethanes and haloacetic acids.

A comparison of the disinfection byproduct formation normalized to the polyelectrolyte concentration is given in Table 3.23. As shown, sunlight combined with permanganate or chlorine caused an increase in the TTHMs and HAA_6 formation potential for Superfloc A-100. For Superfloc N-300, exposure to a single oxidant or sunlight combined with an oxidant increased the disinfection byproduct formation potential. Conversely, a decrease in trihalomethane formation potential for Superfloc A-130 resulted from exposure sunlight and/or chlorine or permanganate. However, an increase in HAA_6 formation potential resulted from exposure to sunlight in combination with either permanganate or chlorine.

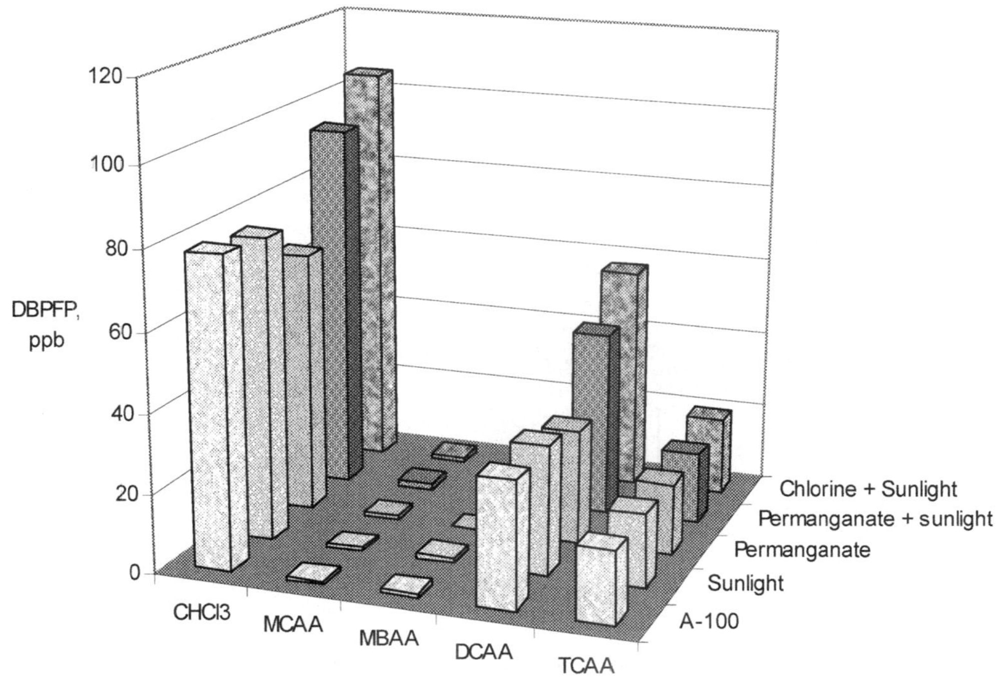

Figure 3-18. Comparison of the impact of sunlight (1 hour exposure), permanganate (0.8 mg/L), chlorine (5 mg/L) and sunlight (1 hour), on the trihalomethane and haloacetic acid formation potential of 5 mg/L of Superfloc A-100.

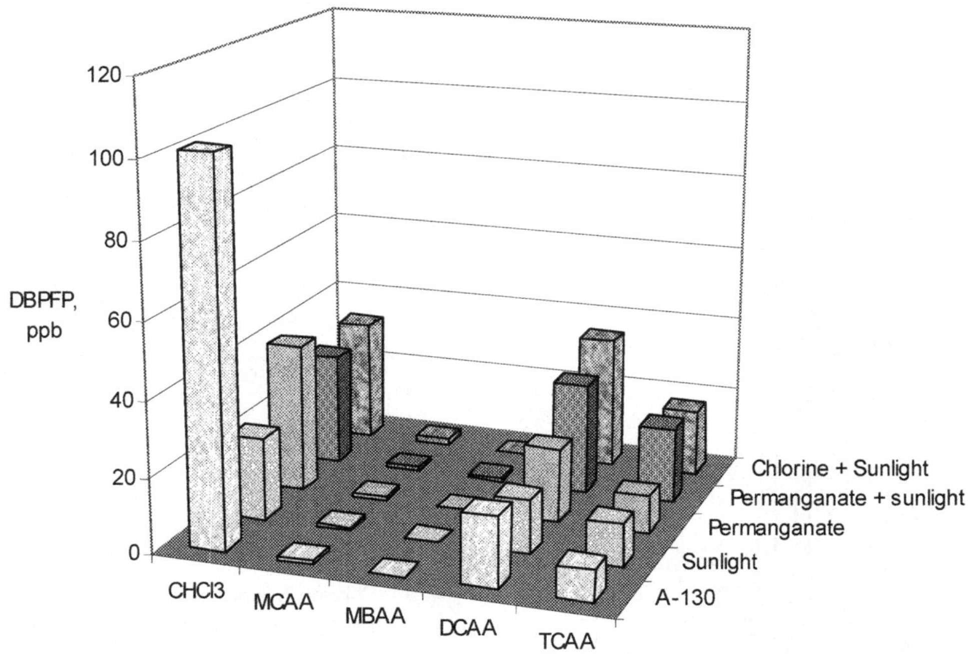

Figure 3-19. Comparison of the impact of sunlight (1 hour exposure), permanganate (0.8 mg/L), chlorine (5 mg/L) and sunlight (1 hour), on the trihalomethane and haloacetic acid formation potential of 5 mg/L of Superfloc A-130.

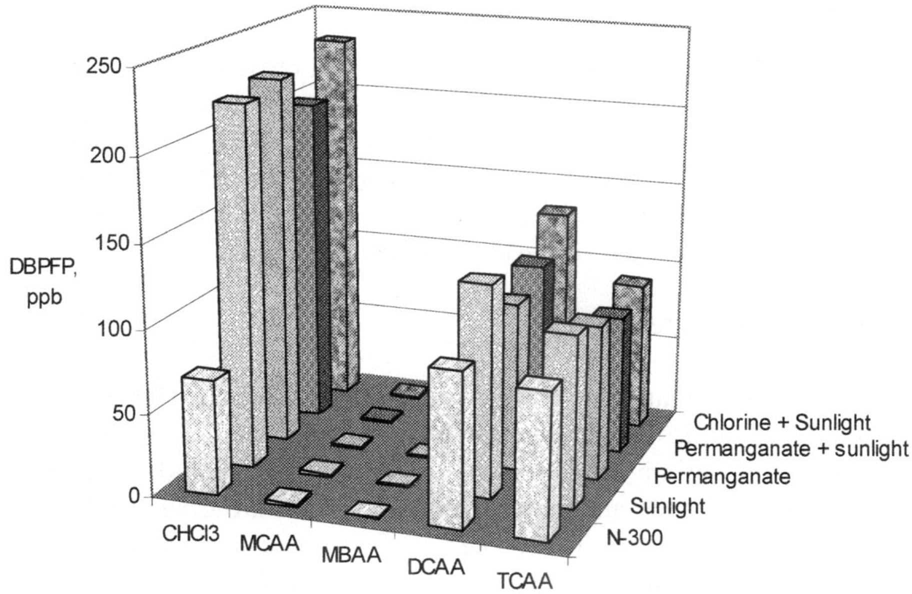

Figure 3-20. Comparison of the impact of sunlight (1 hour exposure), permanganate (0.8 mg/L), chlorine (5 mg/L) and sunlight (1 hour), on the trihalomethane and haloacetic acid formation potential of 5 mg/L of Superfloc N-300.

Table 3-23
Comparison of the Disinfection byproduct formation potential associated with pre-exposure of polyelectrolyte solutions to oxidant chemicals and/or sunlight.

	Disinfection byproduct formation potential, μg DBP/ mg polyelectrolyte				
Polyelectrolyte, DBP	No pre-treatment	Sunlight, 1 hour exposure	Permanganate 0.8 mg/L	Sunlight + Permanganate, 0.8 mg/L	Sunlight (1 hour) + Chlorine, 5 mg/L
Cytec A-100					
TTHM	16	15	13	19	21
HAA_6	11	11	10	13	16
Cytec A-130					
TTHM	20	4	8	6	7
HAA_6	6	5	6	10	11
Cytec N-300					
TTHM	14	44	45	40	47
HAA_6	36	46	39	39	45

PRECOATING OF FILTER MEDIA WITH POLYELECTROLYTES

Filter sand is a variant of silica and has an isoelectric point close to pH 2. Thus, under the pH range of natural waters, sand is negatively charged. Impurities in water such as turbidity and humic materials are also negatively charged. As a result there is a lack of electrostatic driving force for impurities to attach to the filter sand. Coagulation with alum as a pretreatment can alter this situation to some degree. The purpose of this component of the project was to test the feasibility of adding polyelectrolyte to alter the surface charge.

Pilot Plant Testing of Precoating Filter Sand

A continuous pilot scale treatment plant was used to test the feasibility of using polyelectrolytes to precoat filter sand. The degree to which polyelectrolyte was adsorbed onto the sand was evaluated using organic carbon measurements, viscosity testing, and streaming potential measurements. Results of mass balance analyses of organic carbon are given in Table 3.24 and results of viscosity testing are given in Table 3.25. TOC measurements were somewhat inconsistent, however, there was a linear relationship between viscosity and concentration for the high molecular weight polyelectrolytes when measured by the Brookfield DV-11+ viscometer.

Table 3-24

Selected TOC results of polyelectrolyte remaining in solution after contact with sand

Sample	Pre-filter	Initial Carbon, mg	Carbon remaining, mg	Carbon removed, %	Carbon removed, mg/g sand	Polymer removed, mg/g sand
Magnafloc E10						
E10-1	Yes	0.119	0.028	76.6	0.0011	0.0021
E10-2	Yes	0.214	0.108	49.3	0.0016	0.0031
E10-3	Yes	0.399	0.179	55.1	0.0037	0.0074
E10-4	Yes	1.81	1.134	37.3	0.0127	0.0251
E10-5	Yes	23.96	18.28	23.6	0.1100	0.2169
Magnafloc LT20						
LT20-1	Yes	0.179	0.041	77.2	0.002	0.0037
LT20-2	Yes	2.29	1.83	19.9	0.0081	0.0157
BETZ 1160x						
BETZ-1	Yes	0.129	0.012	90.3	0.0012	0.0021
BETZ-2	Yes	1.95	0.54	72.1	0.0269	0.0485
BETZ-1A	No	0.112	0.052	53.6	0.0004	0.0007
BETZ-2A	No	2.07	0.96	53.6	0.0107	0.0194
BETZ-3A	No	18.06	16.40	9.19	0.0316	0.0569
BETZ-1B#	Yes	0.136	0.079	41.5	0.0006	0.0011
BETZ-2B#	Yes	2.054	1.46	28.9	0.0113	0.0203
BETZ-3B#	Yes	19.88	18.55	6.6	0.0258	0.0465
BETZ-1C#	No	0.184	0.10	44.5	0.0012	0.0021
BETZ-2C#	No	2.11	1.66	21.4	0.0085	0.0153
BETZ-3C#	No	20.77	19.95	3.93	0.0156	0.0281

Table 3-25

Supernatant viscosities after sand treatment and calculated loading

Polyelectrolyte	Initial concentration, g/L	Percent removal	Amount adsorbed, mg/g sand, with estimated error
Magnafloc E10	2.753	3.3	0.18±0.8
	0.275	47	0.26±0.5
Magnafloc LT20	2.75	3.6	0.2±0.1
Betz 1160x	2.75	20	1.08±0.1
	0.55	55	0.6±0.05

Based on these results, it appears that the cationic Betz 1160x adsorbed to a greater degree than the non-ionic and anionic polyacryamides. Only a small degree of adsorption was associated with the non-ionic sample Magnafloc LT20 and it was not used in the pilot plant testing.

Results of streaming potential measurements are given in Table 3.26. It is evident from the data that, even at relatively low concentrations of the polyelectrolyte, the surface potential of the sand was influenced by the adsorbed polyelectrolyte. In general, the streaming potential was reduced during treatment again following backwashing. However, it is not known if the reduction in charge was the result of polyelectrolyte being leached from the sand or the effect of the alum floc laden with anionic NOM adsorbing on top of the polyelectrolyte layer.

Table 3-26
Streaming Potentials of Polymer Pretreated Sand Standards

Sample	Polymer Dose (mg/g sand)	Potential (mV) before use	Potential (mV) after use	Potential (mV) after backwash
Washed Sand		-28.4 ± 0.8		-24.5±0.8
	0.0	-28.7± 1.8		-31.3±0.6
Magnafloc E10	0.01	-29.3 ± 0.7		
(Very high MW,	0.05			-28.4±0.7
Low Anionic charge)	0.1	-22.2 ± 1.0		-28.3±1.5
	1.0	-25.1 ± 0.7		-25.3±1.0
	10.0	-25.3 ± 1.6		
Magnafloc LT 20	0.01	-23.6 ± 0.6		
(Non-ionic)	0.1	-15.1 ± 0.6		
	1.0	-9.7 ± 0.4		
	10.0	-8.9 ± 0.4 (14/2/00) -8.9 ± 0.6 (16/2/00)		
BETZ 1160x	0.01	-19.1 ± 1.5		-26.5± 2.0
(Med.- High MW,	0.05			-23.4±1.5
10% Cationic)	0.1	-16.0 ± 0.8		
	1.0	-6.9 ± 0.9		-24.0±0.5
	10.0	-3.9 ± 1.3		
Magnafloc LT35	0.01	-18.3 ± 1.1		
(Med. MW,	0.05	-4.0±1.5	-14.2	
100% Cationic charge)	0.1	+2.6 ± 1.1		
	1.0	+17.9 ± 1.3		-15.8±0.5,
	10.0	+17.7 ± 1.4		-13.8±5.1

Results of Pilot Plant Testing

The pilot plant tests were conducted with dosages of polyelectrolyte ranging from 0 to 1 mg/g of sand. Sand was preconditioned with polyelectrolyte at the rate of 1mg poly /g sand. During these tests, the turbidity of the source water was relatively low (2-3 NTU) and the run length before breakthrough with untreated sand was around 6 hours. When the sand was precoated with Magnafloc E10, the filter run time was over 11 hours for the initial run. However, in subsequent testing, the run time was similar to that of the uncoated sand as shown in Figure 3.21. Polyelectrolyte treatment of the sand had only a marginal effect on the length of time required for filter ripening. Similar results were obtained for Betz 1160x as shown in Figure 3.22. These tests demonstrate that cationic or anionic polyelectrolytes appear to have a similar effect in the initial cycle. In contrast, precoating of the sand with LT35 resulted in a slight increase in the run length in the first two cycles and more rapid filter ripening as shown in Figure 3.23.

At lower dosages of polyelectrolyte, the effect of precoating the media was reduced. All three polyelectrolytes tested were capable of prolonging the filter run for about an additional hour of run time at a level of 0.05 mg /g sand but this effect lasted only one cycle. For a sand bed of 2.75 kg, 138 mg of polyelectrolyte would be required to produce an additional 30L of water per cycle. This represents a polyelectrolyte dose of 4.6 mg/L in addition to the alum dose of 60 mg/L to produce the additional water. Based on this analysis, it appears that precoating the filter media is not economical in comparison with direct dosing of an appropriate polyelectrolyte to the water during clarification.

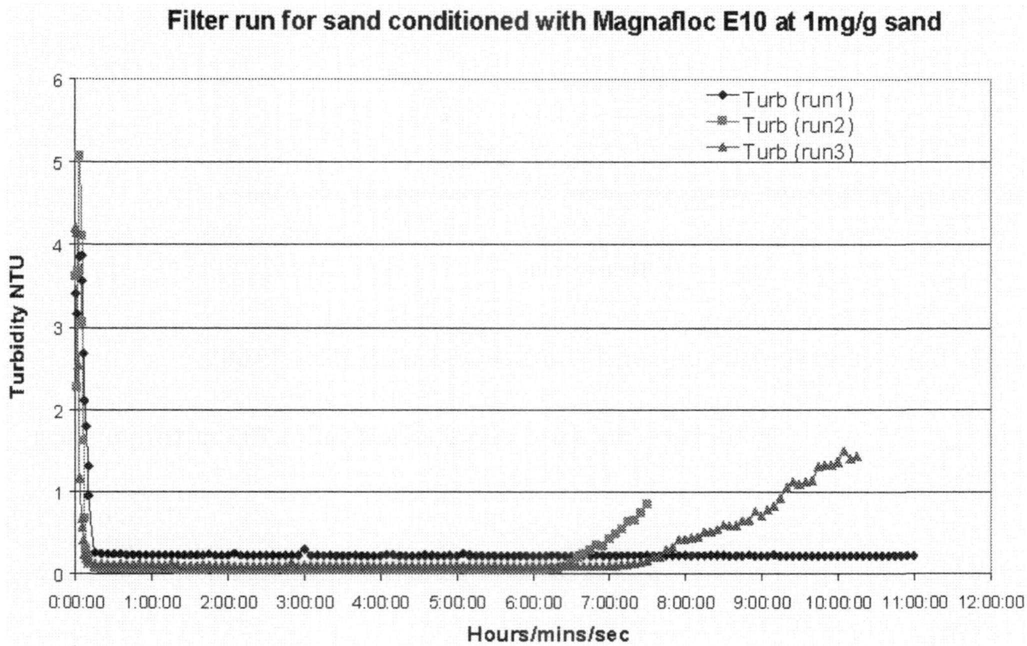

Figure 3-21. Consecutive runs on Magnafloc E10 treated sand.

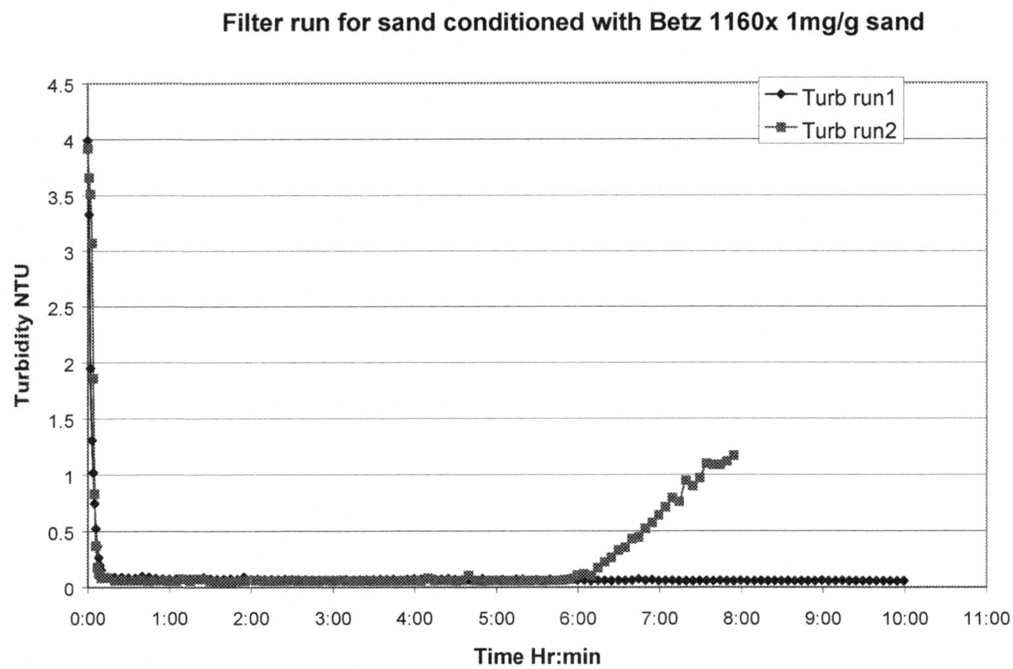

Figure 3-22. Consecutive runs on sand conditioned with Betz 1160x

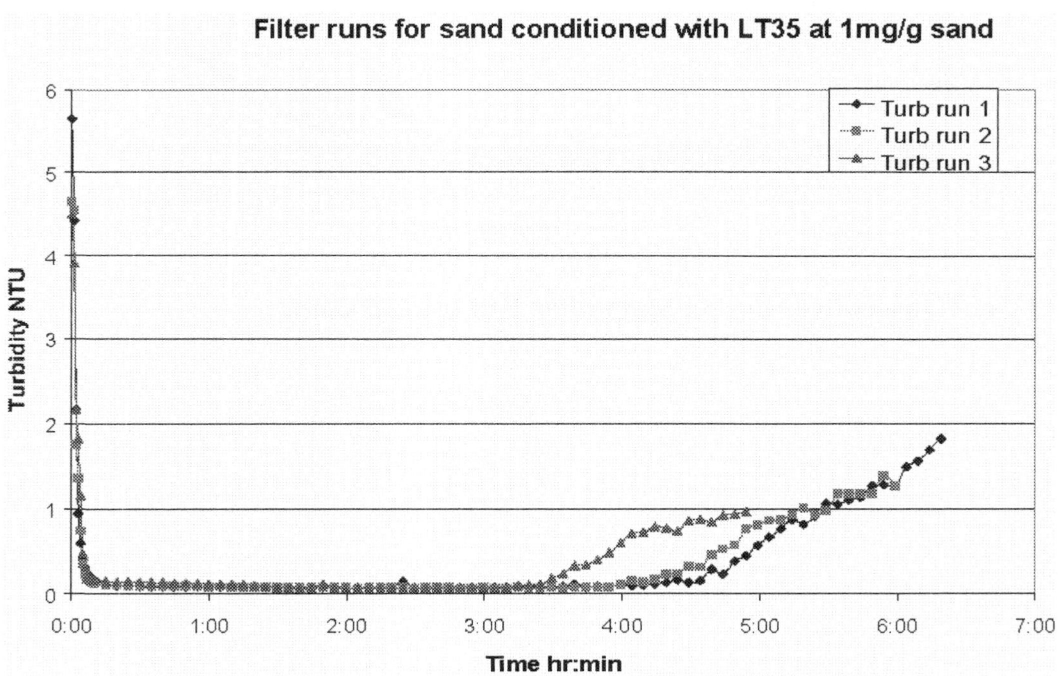

Figure 3-23. Consecutive runs an LT35 treated sand but with higher turbidity water than was used with Magnafloc E10 or Betz 1160x

Pilot Plant Backwashing

The turbidity profile of the backwash was followed for each run. For the precoated sand, the backwash profile of the first cycle was lower and broader than subsequent cycles. The backwash for the first cycle required almost twice the volume of water as the uncoated media. The release of the floc from the sand bed appeared to be inhibited in the first cycle. The size of the flocs in the first backwash appeared larger by visual inspection and may have resulted in misleading turbidity measurements. An example of the backwash profiles is shown in Figure 3.24. While it appears that some benefits are associated with polyelectrolyte conditioned sand in terms of the run time, the backwashing process is much less efficient. The backwash water from the polyelectrolyte treated sand formed flocs more readily than for the untreated sand indicating that residual polyelectrolyte was probably present. The presence of the residual polyelectrolyte could be beneficial in improving treatment of the backwash water.

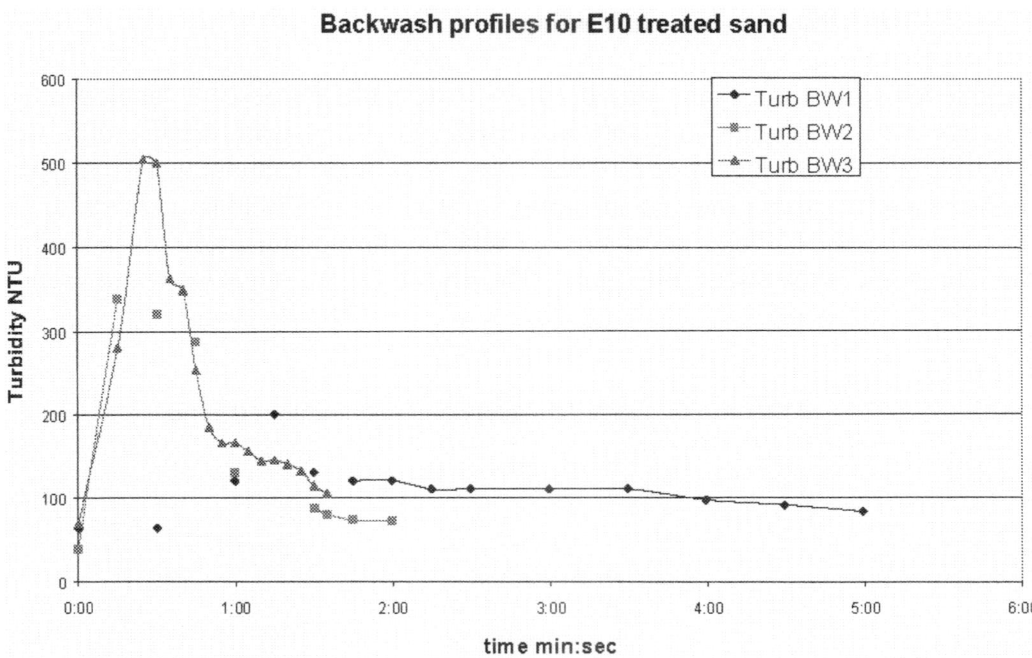

Figure 3-24. Backwash profiles for Magnafloc E10 treated sand at 0.1 mg/g sand.

Mass balances

The mass of solids recovered from backwashing were measured for each run. For an alum dose rate of 60 mg/L (5.14 mg Al/L),with iron removal from the raw water of 1.2 mg/L, turbidity removal of 7 NTU and color removal of 50 PCU, and using the formula of Hall and Hyde (1992) it is possible to calculate the mass of sludge solids as follows:

Sludge solids mg/L	=2x turbidity removed	=14 mg/L
	+ 0.2 colour removed	=10 mg/L
	+ 2.9 Al precipitated	=15 mg/L
	+ 1.9 Fe precipitated	= 2 mg/L
	TOTAL	=41 mg/L

Thus for the 120 L water treated over a 4 hour run, 4.92 g solids were produced and recovered either in the clarifier or the sand filter. Usually 60 – 65% was recovered in the clarifier and 35 – 40% was recovered from the backwash. In most cases where the sand was polymer conditioned, fewer solids were recovered in the backwash from the first cycle than the subsequent cycles; however due to the large variability in results no definite conclusions can be drawn and no trends due to polyelectrolyte dose or type can be discerned.

Fluorescent Labeled Polyacrylamide Testing

A smaller scale pilot plant was used for to test the fluorescent labeled polyelectrolyte. Approximately 1/6 the amount of sand was used. A fluorescent version of a 10% charged polyacrylamide similar to Magnafloc E10 was used. A solution containing 16.6 mg was used with 250 g washed sand. Analysis showed adsorption of 0.0138 mg/g sand (equivalent to 21.7%). The sand was then used for a filter run where the relative throughput (m/hr) was around ten times higher than that used in the larger pilot plant. The profile for the run is shown in Figure 3.25.

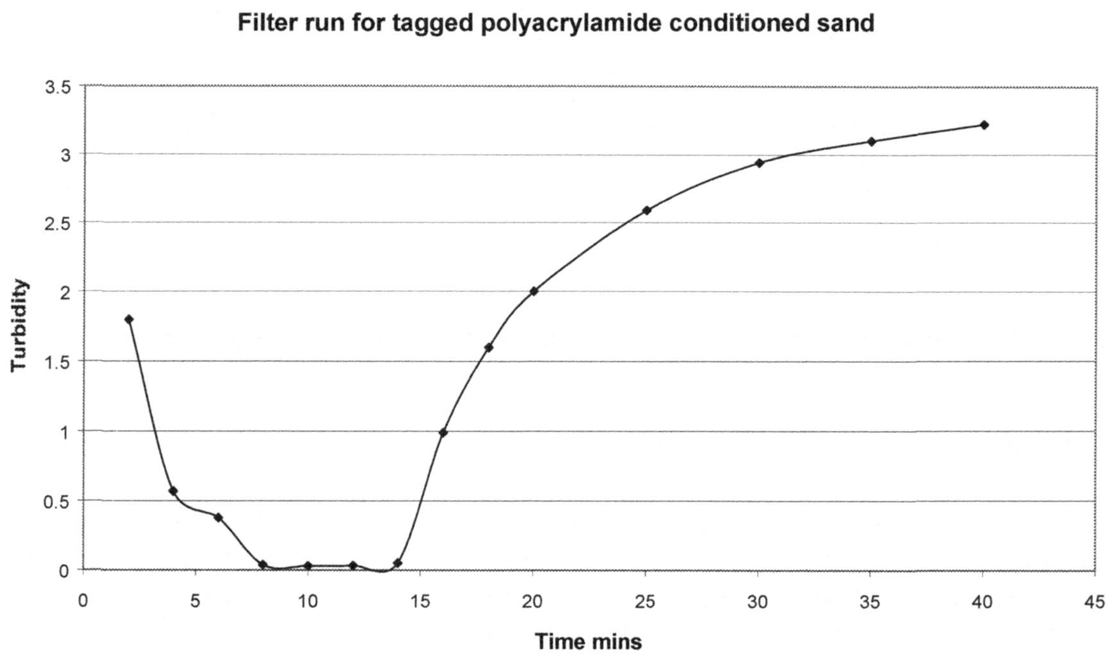

Figure 3-25. Filter run for tagged polyacrylamide conditioned sand – cycle 1.

A mass balance analysis of fluorescently labeled polyelectrolyte is shown in Table 3.27. When sand is coated with a high molecular weight 10% charged polyacrylamide at about 0.0138 mg/g sand, about 20% is loosely bound and is removed by the initial exposure to water. Another ~35% is removed during the first cycle of filtration and backwash. Analysis of what remains on the sand at a moderate level of shear indicates that part of the remainder is capable of being removed under these conditions. This leaves ~35% still on the surface of the sand. With even more severe stripping conditions more polyelectrolyte is removed but the turbidity formed/released masks the analysis. These results help to explain the behavior observed during the pilot plant tests in that the polyelectrolyte adsorbed layer is not stable under filtration or backwashing conditions.

Table 3-27
Fluorescence analysis of filtration, backwash and stripping fractions

Stage of filtration test	Polyelectrolyte Recovery mg	Mass Balance
Loading		
Adsorbed to sand	3.6	3.6
Filter run		
Wash prior to run	0.69	
Composite samples of product water	0.43	
Backwash	0.58	1.7
First stripping of sand	0.70	
Second stripping of sand	0.5 with high uncertainty due to turbidity present	~1.2

These results might reflect the mechanism by which polyelectrolyte adsorbs to the sand surface. When the sand is initially washed some of the less entangled chains of polyelectrolyte are released from the surface as there is no favorable electrostatic charge to attract them. During filtration, alum flocs of high surface area containing color and turbidity attach to the polyelectrolyte. The higher shear conditions that exist in backwashing remove the polyelectrolyte that is attached to the alum flocs.

In summary, pretreatment of the filter media with the three polyacrylamides caused a significant improvement in the first filter run (as monitored by run length) but not in subsequent runs (these closely resembled the runs with untreated sand). Conversely backwashing took longer and generated more backwash water, to reach the same target turbidity level. It is interesting to note that there was very little difference between the effects of oppositely charged polyacrylamides.

CHAPTER 4: CONCLUSIONS AND RECOMMENDATIONS FOR FURTHER RESEARCH

This project was conducted to gain a better understanding of the potential for polyelectrolyte interactions to occur with other treatment chemicals within water treatment facilities. The project involved conduct of bench-scale and pilot-scale testing and analysis of current chemical use practices at water treatment facilities.

CONCLUSIONS

The major conclusions of this project are:

1. The dominant types of polyelectrolytes used in production of drinking water are polyacrylamides, polyDADMACs, and EPI-DMAs. Other chemicals that are used concurrently with polyelectrolytes include oxidants (permanganate, chlorine, ozone, chlorine dioxide), inorganic coagulants, lime, and corrosion inhibitors.
2. When polyelectrolytes are exposed to ultraviolet light in the presence of oxidant chemicals, advanced oxidation reactions can occur. These advanced oxidation reactions can cause decomposition of the polyelectrolyte resulting in reduced performance.
3. Evidence exists that some combinations of polyelectrolytes, chlorine, and ultraviolet light can result in the formation of disinfection byproducts. The levels of regulated disinfection byproducts (trihalomethanes and haloacetic acids) formed by polyelectrolyte reactions, when using typical polyelectrolyte dosages, can range from 4 to 47 μg DBP/mg polyelectrolyte.
4. The use of cationic or anionic polyelectrolytes of low charge density to precoat filter media can result in temporary increases of one to five hours in filter run times. Precoating filter media with a cationic polyelectrolyte of medium molecular weight and high charge density (1 mg/g) appeared to result in about a 20% decrease in the filter ripening period. However, the effects of precoating do not appear to be sustainable over more than three filter runs. Based on the quantities of polyelectrolytes needed for precoating the filter media and the lack of sustainability, this process does not appear to be cost-effective for conventional filter applications.
5. Fluorescent labels can be attached to polyelectrolytes to allow for the use of fluorescence monitoring. The preliminary tests conducted in this study suggest that further refinement of the fluorescent labeling technique could provide a useful tool for rapid monitoring of residual polyelectrolytes. The ability to monitor residual polyelectrolytes would help to improve polyelectrolyte dosage control and process performance.

RECOMMENDATIONS FOR FURTHER RESEARCH

Because of the wide variety of potential chemical interactions that can occur in water treatment facilities, additional research is needed to assess the degree to which these reactions occur. Some suggested research topics are provided below:

1. **Monitoring tools**. The low levels of residual polyelectrolytes in treated water impose analytical challenges for process monitoring. Development of improved tools for quantifying residual polyelectrolytes, such as the use of fluorescent labels, antibody based analytical methods, or on-line sensors may be valuable in optimization of chemical dosages and reducing the potential for chemical interactions to occur.
2. **UV Disinfection**. With increasing interest in the use of UV disinfection, it is important to continue to assess potential interactions between polyelectrolytes and other chemicals (naturally occurring and added in treatment) and UV irradiation. Depending on the concentration of polyelectrolyte and the UV fluence, the interactions could include partial decomposition of polyelectrolytes and fouling of UV lamps, and/or contribution to formation of disinfection byproducts. Testing should be conducted over a range of water quality conditions and treatment applications to elucidate these reactions.
3. **Membrane Processes.** To date there is limited information on the impacts of polyelectrolytes on membrane processes. Frequently, pretreatment for membrane systems includes coagulation/flocculation/filtration for removal of particulates. In addition, oxidizing and reducing agents may be added upstream of membranes. Residual polyelectrolytes have the potential to cause problems with membrane operation due to fouling of membrane surfaces. In addition, partially decomposed polyelectrolytes could serve as disinfection byproduct precursors or provide a substrate for microorganisms that contribute to biofouling. If polyelectrolytes are used upstream of membrane systems, it is important to understand the potential reactions that can occur and to implement measures to minimize these problems.
4. **Recirculated flows .** It is fairly common practice to recirculate flows from filter backwash water, gravity thickeners, and belt filter presses. Recirculated flows consist of supernatant (i.e. from sludge thickeners) and filtrates (from filter presses). In many cases, these return flows contain residual polyelectrolytes either from the original flowstream or due to addition during residuals processing (filter backwash treatment, filter presses). It is important to optimize the approach used for managing the recirculated flows to prevent reactions between residual polyelectrolytes and other chemicals that are added to the flowstream.

CHAPTER 5: RECOMMENDATIONS TO UTILITIES

The simultaneous use of multiple chemicals is widespread in the drinking water industry. While each chemical is employed for a specific purpose, all of the chemicals that are used have the potential to interact within the water matrix. The net effect of chemical interactions can be positive, negative, or neutral; depending on water quality. This research project has provided a framework for evaluating the extent of chemical interactions that occur during drinking water treatment.

Chemical costs represent a major operating expense for water utilities. Polyelectrolytes and oxidant chemicals are the some of the most costly chemicals. Through this project, it is clear that the concurrent presence of oxidant chemicals and polyelectrolytes can result in degradation of both types of chemicals, thereby decreasing process efficiency and introducing unnecessary operating costs.

Exposure to sunlight can introduce an additional oxidant demand on water and can also result in advanced oxidation reactions. When polyelectrolytes are added to water, either as part of treatment or through recycled flows, ultraviolet light can react with the polyelectrolytes, potentially compromising process performance. The advanced oxidation effects were observed to different degrees for polyacrylamides, polyDADMACs and EPI-DMAs, with the extent of reaction specific to the type of polyelectrolyte and water quality parameters.

There are six important recommendations to utilities that can be derived from this project.

1. **Use of polyelectrolytes in conjunction with other chemicals.**
   Polyelectrolytes are effective for improving the efficiency of coagulation, flocculation, ballasted sedimentation, and filtration. In the design and operation of chemical feed facilities, it is important that adequate contact time be available to segregate the addition of polyelectrolytes from addition of other chemicals, particularly oxidant chemicals (i.e. permanganate, chlorine, ozone) to prevent carry-over of residual polyelectrolyte into downstream processes.

2. **Make-up water for polyelectrolyte addition**.
   Typically, polyelectrolytes are supplied in either dry form or as concentrated solutions. Prior to injection into the treatment train, dry chemicals are hydrated or dissolved. To optimize the effectiveness of polyelectrolytes, it is important that the make-up water is free of excessive levels of oxidants (i.e. chlorine residual or permanganate). Usually the most cost-effective make-up water is finished water, however, finished water frequently contains a disinfectant residual. It is suggested that preliminary tests be conducted to ensure the compatibility of the make-up water with the polylectrolyte solution.

3. **Indirect monitoring tools for polyelectrolyte application optimization**.

Due to the relatively low concentrations of polyelectrolytes typically used in potable water treatment, direct monitoring of residuals is not practical using currently available monitoring technology. Indirect monitoring tools should be used to help ensure that polyelectrolyte dosing and mixing conditions are appropriate. Examples of indirect monitoring tools include streaming current, residual TOC, or TOC removal efficiency (particularly for waters with TOC levels below 2 mg/L), disinfection byproduct formation potential (and formation potential removal), oxidant demand, and UV_{220}.

Routine monitoring should be conducted immediately after polyelectrolyte addition and downstream of all locations where polyelectrolytes are reacting and being removed (i.e. flocculation, sedimentation, or filtration). If polyelectrolytes are overdosed, or dosed too close to other chemical application points, potential reactions may occur with other chemicals, particularly in the presence of natural or applied UV light. Such reactions can result in the formation of disinfection byproducts or byproduct precursors and can also result in chemical fouling of filter media and UV lamps.

4. **Use of polyelectrolytes in reaction basins exposed to sunlight.**

Some types of polyelectrolytes decompose due to exposure to UV light. The decomposition reactions are exacerbated by the presence of chemical oxidants such as chlorine or permanganate. If possible, polyelectrolyte addition and mixing should be completed prior to sunlight exposure. Similar precautions should be exercised if polyelectrolytes are used upstream of UV disinfection processes.

5. **Recirculation of solids containing unreacted polyelectrolytes and reacted polyelectrolytes**.

In many water treatment facilities, residual solids are recirculated within the plant either from filter backwashing recycle streams, return flows from thickening basins, or via solids maintained intentionally in solids-contact upflow clarifiers. Residual solids may contain residual polyelectrolytes providing a mechanism for inadvertent introduction of polyelectrolytes back into the flowstream. When these concentrated zones of polyelectrolytes are exposed to other treatment chemicals and/or UV, the potential for chemical interactions can increase. It is suggested that steps be taken to avoid this problem in the development of dosing strategies for polyelectrolytes and other treatment chemicals. Design and operation of systems for solids removal and recirculation of return flows should also consider the potential impacts of polyelectrolyte interactions.

6. **Emerging health effects concerns**.

Currently, the Safe Drinking Water Act (SDWA) includes limits on the allowable concentrations of acrylamide and epichlorohydrin monomers in drinking water supplies. Recent research (AwwaRF 2678) and the genesis of revised AWWA standards have demonstrated that commonly used polyelectrolytes such as epichlorhydrin-dimethylamine (EPI-DMA) and poly(Diallydimethylammonium chloride (polyDADMAC) may react with disinfectants to form significant levels of N-nitrosodimethylamine (NDMA), a suspected carcinogen with a theoretical lifetime cancer risk level of 10^{-6} from exposures levels of 0.7 ng/L in drinking water (Valentine et al. 2004). This project helps to inform utilities about the potential for undesirable side-reactions associated with polyelectrolyte use and provides advice that can be applied towards design and operations concerns.

APPENDIX: COMMONLY USED POLYELECTROLYTES

Polyelectrolyte Suppliers

Company	**Address**	**Phone, Fax**	**Web Site**
BetzDearborn	4356 Somerton Road Trevose, PA 19053-6783	215 355 3300 215 953 5524	www.betzdearborn.com
Buckman Laboratories International Inc.	1256 North Mclean Blvd Memphis, TN 38108-0305	901 278 0330 901 276 5343	www.buckman.com
Ciba Specialty Chemicals	2301 Wilroy Road Suffolk, VA 23439-0820	757 538 5321 757 538 3989	www.cibasc.com
CYTEC Industries Inc.	5 Garret Mountain Plaza W. Paterson, NJ 07424	973 357 3100 973 357 3078	www.cytec.com
ONDEO Nalco Company	1601 W. Diehl Road Naperville, IL 60563-1198	630 305 1000 630 305 2900	www.ondeonalco.com
Polydyne, Inc.	1 Chemical Plant Rd. Riceboro, GA 31323	800-848-7659 912-880-2044	www.polydyneinc.com
SNF Floerger	Pearl River Polymers Inc. 1-59 North Exit 5A Pearl River, LA 70452	504 863 5703 504 863 7580	www.floerger.com
Stockhausen Inc.	2401 Doyle Street Greensboro , NC 27406	336 333 3534 336 379 3564	www.stockhausen.com

Details of Commercially Available Polyelectrolytes

Dose limits are from the US National Sanitation Foundation web site at www.nsf.org. Polymers that have been given a rating by NSF at the time of publication are marked in bold type in all the Tables in this Appendix.

Betzdearborn Inc.

Polymer Type	Commercial Name	Form	MW	CD	NSF Max. Dose, mg/L
polyDADMAC	Klaraid PC1192P	solution	M	VH	50
EPI-DMA	Klaraid PC1182	solution	L	VH	
'Polyamines'	Klaraid PC1195P	solution	L	VH	6.4
	Klaraid PC1190P	solution	L	VH	24
CPAM	Polyfloc CP1158	powder	M	H	
	Polyfloc CP1160P	powder	H	L	1
	Polyfloc CE1167	emulsion	H	L	
APAM	Polyfloc AP1100P	powder	H	L	1
	Polyfloc AP1110P	powder	H	M	1
	Polyfloc AP1115P	emulsion	H	L	1
	Polyfloc AP1120P	powder	H	H	1
	Polyfloc AP1138	powder	VH	H	
PAM	Polyfloc AP1142	powder	H	0	

Buckman Laboratories International Inc.

Polymer Type	Commercial Name	CD, mol%	NSF Max. Dose, mg/L
polyDADMAC	Bufloc 584	100	
EPI-DMA	Bulab 5031P	100	20
'Polyamines'	Bulab 5132P	100	20
Polyacrylamides	Bulab 5112P		1
	Bulab 5196P		1
	Bulab 5229P		1
	Bulab 5273P		1

Ciba Specialty Chemicals

Polymer Type	Commercial Name	Form	MW	CD, mol%	NSF Max. Dose, mg/L
polyDADMAC	Magnafloc LT-7985	solution	L	100	50
	Magnafloc LT-7992	solution	H	100	50
	Magnafloc LT-7993	solution	H	100	50
	Magnafloc LT-7994	solution	M	100	25
	Magnafloc LT-7995	solution	L	100	25
	Magnafloc LT-7996	solution	H	100	50
	Magnafloc LT-7998	solution	L	100	29
EPI-DMA	Magnafloc LT-7981	solution	L	100	20
'Polyamines'	Magnafloc LT-7990	solution	M	100	20
	Magnafloc LT-7991	solution	H	100	20
CPAM	Magnafloc LT 7922	emulsion	VH	L	
	Magnafloc LT 22	granules	M	L	1
	Magnafloc LT 22S	granules	H	L	1
	Magnafloc LT 24	granules	M	VL	1
	Zetag 7501	microbeads	L	VH	
	Zetag 7503	microbeads	M	H	
	Zetag 7504	microbeads	L	H	
	Zetag 7529	microbeads	L	L	
	Zetag 7547	microbeads	M	L	
	Zetag 7553	microbeads	H	H	
	Zetag 7555	microbeads	M	H	
	Zetag 7557	microbeads	M	H	
	Zetag 7563	microbeads	M	M	
	Zetag 7565	microbeads	VH	L	
	Zetag 7566	microbeads	M	M	
	Zetag 7565	microbeads	VH	L	
	Zetag 7587	microbeads	M	H	
	Zetag 7623	granules	VH	VL	
	Zetag 7630	granules	VH	L	
	Zetag 7631	granules	M	H	
	Zetag 7633	granules	M	H	

Polymer Type	Commercial Name	Form	MW	CD, mol%	NSF Max. Dose, mg/L
CPAM	Zetag 7634	granules	VH	VL	
	Zetag 7635	granules	M	H	
	Zetag 7645	granules	VH	VL	
	Zetag 7648	granules	H	L	
	Zetag 7650	granules	VH	L	
	Zetag 7651	granules	VH	H	
	Zetag 7652	granules	VH	H	
	Zetag 7653	granules	VH	M	
	Zetag 7654	granules	VH	VH	
	Zetag 7664	granules	H	M	
CPAM	Zetag 7689	granules	H	VH	
	Zetag 7692	granules	H	L	
	Zetag 7818	emulsion	VH	VH	
	Zetag 7820	emulsion	VH	VL	
	Zetag 7821	emulsion	VH	VL	
	Zetag 7822	emulsion	VH	VL	
	Zetag 7823	emulsion	H	M	
	Zetag 7824	emulsion	H	H	
	Zetag 7824 FS40	emulsion	M	H	
	Zetag 7826	emulsion	H	VH	
	Zetag 7826 FS40	emulsion	M	VH	
	Zetag 7833	emulsion	VH	M	
	Zetag 7834	emulsion	H	H	
	Zetag 7848	emulsion	M	VH	
	Zetag 7867	dispersion	H	M	
	Zetag 7867 FS40	dispersion	L	M	
	Zetag 7869	dispersion	M	VL	
	Zetag 7873	dispersion	H	L	
	Zetag 7875	dispersion	H	H	
	Zetag 7875 FS25	dispersion	M	H	
	Zetag 7875 FS40	dispersion	L	H	
	Zetag 7878	dispersion	H	VH	
	Zetag 7878 FS25	dispersion	M	VH	

Polymer Type	Commercial Name	Form	MW	CD, mol%	NSF Max. Dose, mg/L
CPAM	Zetag 7878 FS40	dispersion	L	VH	
	Zetag 7879	dispersion	H	VH	
	Zetag 7879 FS40	dispersion	M	VH	
	Zetag 7888	dispersion	M	VH	
APAM	Alclar 661	granules	H	VH	
	Alclar 662	granules	VH	VH	
	Magnafloc 10	granules	M	VL	
	Magnafloc 1011	granules	H	M	
	Magnafloc 110L	dispersion	M	M	
	Magnafloc 120L	dispersion	M	H	
	Magnafloc 155	granules	M	L	
	Magnafloc 156	microbeads	M	M	
	Magnafloc 24	microbeads	L	VL	
	Magnafloc 336	granules	VH	M	
	Magnafloc 338	granules	H	VL	
	Magnafloc 340	granules	H	H	
	Magnafloc 342	granules	VH	L	
	Magnafloc 356	granules	M	M	
	Magnafloc 358	granules	M	H	
	Magnafloc 611	granules	M	VH	
APAM	Magnafloc 90L	dispersion	M	L	
	Magnafloc 919	granules	H	H	
	Magnafloc E 32	emulsion	H	VL	1
	Magnafloc E 34	emulsion	VH	L	
	Magnafloc E 38	emulsion	VH	H	1
	Magnafloc E 40	emulsion	VH	H	
	Magnafloc E 42	emulsion	VH	H	
	Magnafloc LT 25	granules	M	L	1
	Magnafloc LT 26	granules	M	M	1
	Magnafloc LT 27	granules	H	M	1
PAM	Magnafloc LT 20	granules	M	0	1
	Magnafloc 333	granules	H	0	
	Magnafloc 351	granules	M	0	
	Magnafloc 80 L	dispersion	L	0	
	Magnafloc E 30	emulsion	H	0	1

CYTEC Industries Inc.

Further details are available from the Cytec web site at www.cytec.com.

Polymer Type	Commercial Name	Form	MW	CD, mol%	NSF Max. Dose, mg/L
polyDADMAC	Superfloc C-587	solution	M	100	50
	Superfloc C-587 HV	solution	M	100	50
	Superfloc C-591	solution	H	100	50
	Superfloc C-592	solution	M	100	25
	Superfloc C-595	solution	VH	100	25
EPI-DMA	Superfloc C-521	solution	M	100	
'Polyamines'	Superfloc C-567	solution	VL	100	7
	Superfloc C-572	solution	VL	100	20
	Superfloc C-573	solution	L	100	20
	Superfloc C-577	solution	M	100	20
	Superfloc C-578	solution	M-H	100	
	Superfloc C-581	solution	H	100	20
CPAM	Superfloc MX 5	emulsion	Ultra H	3	
	Superfloc MX 10	emulsion	Ultra H	5	
	Superfloc MX 20	emulsion	Ultra H	10	
	Superfloc MX 40	emulsion	Ultra H	20	
	Superfloc MX 45	emulsion	Ultra H	20	
	Superfloc MX 60	emulsion	Ultra H	40	
	Superfloc MX 80	emulsion	Ultra H	55	
	Superfloc SD 2060	emulsion	H	40	
	Superfloc SD 2061	emulsion	H	40	
	Superfloc SD 2065	emulsion	M	35	
	Superfloc SD 2080	emulsion	H	55	
	Superfloc SD 2081	emulsion	H	55	
	Superfloc SD 2083	emulsion	M-H	55	
	Superfloc SD 2085	emulsion	M	55	
	Superfloc C-491	powder	H	5	
	Superfloc C-491 K	powder	H	3	
	Superfloc C-492	powder	H	10	1
	Superfloc C-492 HMW	powder	VH	40	1
	Superfloc C-494	powder	H	20	
	Superfloc C-494 HMW	powder	H	20	
	Superfloc C-496	powder	H	35	1
	Superfloc C-496 HMW	powder	H	35	1

Polymer Type	Commercial Name	Form	MW	CD, mol%	NSF Max. Dose, mg/L
CPAM	Superfloc C-498	powder	H	55	
	Superfloc C-498 HMW	powder	H	55	
	Superfloc C-442	powder	M	10	
	Superfloc C-444	powder	M	20	
	Superfloc C-448	powder	M	55	
	Superfloc C-1555	emulsion	H	2.5	
CPAM	Superfloc C-1590	emulsion	H	1	
	Superfloc C-446	powder	M	35	
	Superfloc C-1592	emulsion	H	10	3.1
	Superfloc C-1594	emulsion	H	20	
	Superfloc C-1596	emulsion	H	40	
	Superfloc C-1598	emulsion	H	55	
	Superfloc C-1598 PG	emulsion	H	55	2.5
	Superfloc C-1596 PG	emulsion	H	40	
	Superfloc C-1502	modified emulsion	H	20	
	Superfloc C-1504	modified emulsion	H	40	
	Superfloc C-1505	modified emulsion	M	55	
	Superfloc C-1500	modified emulsion	M	80	
	Superfloc C-1507	modified emulsion	M	80	
Mannich Bases	Superfloc C-2535H	solution	H	40	
	Superfloc C-2540	solution	M	55	
APAM	Superfloc A-100	powder	H	7	2.5
	Superfloc A-110	powder	H	18	3
	Superfloc A-130	powder	H	35	1
	Superfloc A-130 HMW	powder	VH	35	1
	Superfloc A-150	powder	VH	50	
	Superfloc A-1849 RS	emulsion	M	3	3.5
	Superfloc A-1820	emulsion	H	30	
	Superfloc A-1883 RS	emulsion	H	30	3.5
	Superfloc A-1885 RS	emulsion	H	30	
	Superfloc A-870	powder	VL	70	
PAM	Superfloc N-300 LMW	powder	M	0	1
	Superfloc N300	powder	H	0	1
	Superfloc N1986	emulsion	M	0	3.5

ONDEO Nalco Company

Polymer Type	Commercial Name	CD,mol%	NSF Max. Dose, mg/L
polyDADMAC	Nalcolyte 8103	100	41.6
	Nalcolyte 8793	100	41.6
	Nalcolyte 8108	100	33.3
	Nalcolyte 8958	100	33.3
	Nalcolyte 8799	100	10
	Nalcolyte 8102	100	15
	Nalcolyte 8792	100	15
	Cat-Floc TL	100	50
	Cat-Floc T	100	50
	Cat-Floc LS	100	100
	Cat-Floc A	100	81
	Cat-Floc T-2	100	23
	Cat-Floc CL	100	25
	Cat-Floc 8799 Plus	100	23
	Cat-Floc 71259	100	23
	Cat-Floc CFCL	100	25
	Cat-Floc 8102 Plus	100	50
	Cat-Floc 8792	100	50
	Cat-Floc CFTL	100	50
	Cat-Floc CFA	100	81
	Cat-Floc EL	100	26
	Cat-Floc DL	100	57
	Cat-Floc 8103 Plus	100	57
	Cat-Floc 8793 Plus	100	57
	Cat-Floc L	100	50
	Cat-Floc 8108 Plus	100	50
	Cat-Floc 8958 Plus	100	50
	Cat-Floc C	100	25
	Cat-Floc 2953	100	25
	Nalco 7000	100	25
	E-905	100	50
	E-2231	100	100
EPI-DMA	Nalcolyte 8100	100	20
'Polyamines'	Nalcolyte 8105	100	20
	Nalcolyte 8790	100	20
	Nalcolyte 8795	100	20
Polyacrylamides	Nalco 1C34		1
	Nalclear 7744		1
	Nalclear 7766 Plus		1

Polymer Type	Commercial Name	CD, mol%	NSF Max. Dose, mg/L
Polyacrylamides	Nalclear 7768		1
	Nalclear 8173 PULV		1
	Nalclear 8181		1
	Nalclear 8182		0.9
	Nalclear 8184		1
	Nalclear 8472		0.9
	Nalclear 8478		1
	Nalclear 8771		1
	Nalclear 8772		1
	Nalclear 8774 Plus		1
	Optimer 7190 Plus		1
	Optimer 8170 PULV		1
	Pol-E-Z 675		3.5
	Pol-E-Z 8736		0.8
	Pol-E-Z 692		0.5
	E-2176		0.8
	E-2279		0.5

Polydyne, Inc.
Subsidiary of SNF Floeger

Polymer Type	Commercial Name	Form	% Actives	Brookfield Viscosity, cPs	MW	CD	NSF Max. Dose, mg/L
Flocculants							
APAM	Clarifloc A-210P	Emulsion	28.4	600-1200		M	1
APAM	Clarifloc A-3333P	Powder				M	1
CPAM	Clarifloc C-3223	Powder				M	1
CPAM	Clarifloc C-3257	Powder				M	1
CPAM	Clarifloc C-3291	Powder				H	
CPAM	Clarifloc C-6240	Emulsion	38.5	<1500		M	1
Nonionic PAM	Clarifloc N-120P	Emulsion	30.3	600-1200		0	1
Nonionic PAM	Clarifloc N-3300P	Powder				0	1
Coagulants							
polyDADMAC	Clarifloc C-308P	Solution	19-21	80-190	L		50
	Clarifloc C-318	Liquid	39-41	1000-3000	L		25
	Clarifloc C-338	Solution	19-21	200-600	M		50
	Clarifloc C-348	Solution	39-41	8000-12,000	M		25
	Clarifloc C-358	Solution	19-21	700-1,000	H		50
	Clarifloc C-368	Solution	34-36	7,000-20,000	H		28
	Clarifloc C-378	Solution	19-21	2,000-3,000	H		50
	Clarifloc LF-3541	Solution	9-11	15-30	L		100
EPI-DMA	Clarifloc C-309P	Solution	48.5-50	650-1,000	M	H	20
	Clarifloc C-339	Solution	48.5-50	250-400	L	H	20
	Clarifloc C-349	Solution	49-51	23-35	L	H	20
	Clarifloc C-359	Solution	49-51	100-200	L	H	
	Clarifloc C-379	Solution	48-50	80-104	L	H	20
	Clarifloc C-391	Solution	48.5-50	5,000-9,000	H	H	20
	Clarifloc C-399	Solution	49-51	1,000-4,000	L	H	20

SNF Floerger

Polymer Type	Commercial Name	Form	% Actives	Brookfield Viscosity, cps	MW	CD	NSF Max. Dose, mg/L
polyDADMAC	Floquat FL 45 DL	solution	19-21	50-150		VH	50
	Floquat FL 45	solution	19-21	200-400		VH	50
	Floquat FL 45 VHM	solution	19-21	600-900		VH	50
	Floquat FL 45 CLV	solution	39-41	1000-3000		VH	25
	Floquat FL 45 C	solution	39-41	8000-13000		VH	25
EPI-DMA	Floquat FL 28 P1	solution	48-51	100-200		VH	20
'Polyamines'	Floquat FL 28 P2	solution	48-51	200-400		VH	
	Floquat FL 28 P3	solution	48-51	300-500		VH	20
	Floquat FL 28 P4	solution	48-51	600-1000		VH	20
	Floquat FL 17	solution	48-51	1000-4000		VH	20
	Floquat FL 18-40	solution	38-41	1300-2200		VH	
CPAM	Flopam FO 4107 PWG	powder	100	95 @ 5 g/L	VH	VL	
	Flopam FO 4115 PWG	powder	100	160	VH	VL	1
	Flopam FO 4140 PWG	powder	100	370	VH	VL	
	Flopam FO 4190 PWG	powder	100	545	VH	L	1
	Flopam FO 4240 PWG	powder	100	615	VH	L	
	Flopam FO 4290 PWG	powder	100	780	VH	M	1
APAM	Flopam AH 912 PWG	powder	100	350 @ 5 g/L	VH	VL	1
	Flopam AN 905 PWG	powder	100	450	VH	VL	1
	Flopam AN 910 PWG	powder	100	800	VH	L	1
	Flopam AN 913 PWG	powder	100	1200	VH	L	
	Flopam AN 923 PWG	powder	100	1395	VH	M	1
	Flopam AN 934 PWG	powder	100	1715	VH	M	1
	Flopam AN 945 PWG	powder	100	1740	VH	H	1
	Flopam AN 956 PWG	powder	100	1600	VH	H	
PAM	Flopam FA 920 PWG	powder	100	100 @ 5 g/L	VH	0	1

Stockhausen GmbH & Co. KG

Polymer Type	Commercial Name	Form	Viscosity, mPas	MW	CD	NSF Max. Dose, mg/L
polyDADMAC	Praestol 186K	solution	< 2500	L	VH	50
	Praestol 186KH	solution	< 2500	L	VH	50
	Praestol 187K	solution	< 2500	L	VH	25
	Sumaclear P30	solution	< 2500	L	VH	25
	Sumaclear P35	solution	< 2500	L	VH	47
EPI-DMA	Praestol 188K	solution	< 2500	L	VH	20
'Polyamines'	Praestol 189K	solution	< 2500	L	VH	20
	Praestol 193K	solution	< 2500	L	VH	20
	Sumaclear P10	solution	< 2500	L	VH	20
	Sumaclear P20	solution	< 2500	L	VH	20
CPAM	Praestol K105L	emulsion	< 4000*	H	L	
	Praestol K110L	emulsion	< 4000*	H	L	1
	Praestol K120L	emulsion	< 4000*	H	L-M	
	Praestol K226FLX	emulsion	< 4500*	H	M	
	Praestol K111L	emulsion	< 4000*	H	M	1
	Praestol K122L	emulsion	< 4000*	H	H	
	Praestol K133L	emulsion	< 4000*	H	H	
	Praestol K136L	emulsion	< 4000*	H	H	
	Praestol K260FL	emulsion	< 5500*	H	VH	
	Praestol K275FLX	emulsion	< 5500*	H	VH	
	Praestol K280FL	emulsion	< 1500*	H	VH	
	Praestol K144L	emulsion	< 4000*	H	VH	
	Praestol K290FL	emulsion	< 1800*	H	VH	
	Praestol K290FLX	emulsion	< 2000*	H	VH	
	Praestol K295FL	emulsion	< 2500*	H	VH	
	Praestol 311 TR	emulsion	> 2000*	H	M	1
	Praestol 321 TR	powder	> 3000*	H	M	1
	Praestol 611 TR	powder	450**	M	L	1
	Praestol 650 TR	powder	700**	M	M	1
	Praestol 851 TR	powder	600**	VH	L	1
	Praestol 852 TR	powder	900**	VH	M	0.4
	Praestol 853 TR	powder	900**	VH	M	0.4
APAM	Praestol 2510	powder	> 300**		L	
	Praestol 2515/2515TR	powder	> 2000**	H	L	1

Polymer Type	Commercial Name	Form	Viscosity, mPas	MW	CD	NSF Max. Dose, mg/L
APAM	Praestol 2520	powder	> 3500**		L-M	
	Praestol 2530/2530TR	powder	> 5000**	H	M	1
	Praestol 2540/2540TR	powder	> 4500**	H	M	1
	Praestol 2640	powder	> 4000**		M	
	Praestol A3010 LTR	emulsion	> 6500**	H	L	1
	Praestol 3050L	emulsion	>10000**		H	
	Praestol 3095L	emulsion	> 5000**		H	
PAM	Praestol 2500/ 2500TR	powder	> 200**	H	0	1
	Praestol N3100 LTR	emulsion	> 300**	H	0	1
	Praestol N3100 KR	emulsion	> 300**	H	VL	1

1 % solution in distilled water; viscosities for the solution polymers are of the products
** 0.5% solution in distilled water

REFERENCES

Aizawa, T., Y. Magara, and M. Musashi. 1991 Problems with introducing synthetic polyelectrolyte coagulants into the water purification process. *Water Supply,* **9,** 27-35.

Aksberg, R. and L. Wagberg. 1989. Hydrolysis of Cationic Polyacrylamides. J. Appl. Pol. Sci. 38, 297-304.

Alekseeva, L. P. and Ya. L. Khromchenko. 1988. Influence of conditions of reagent treatment of water on the formation of chloroform. Soviet J. Water Chem. & Tech. 10, 140-144.

Alt, C. 1989. Centrifuges for sludge treatment. In Water, Wastewater, and Sludge Filtration. Edited by S. Vigneswaran and R. Ben Aim. P. 249, Boca Raton:CRC Press.

American Water Works Association. 1998. AWWA Standard for poly(diallydimethylammonium chloride), ANSI/AWWA B451-98, AWWA. Denver, CO.

American Water Works Association. 1999. AWWA Standard for EPI-DMA Polyamines, ANSI/AWWA B452-98, AWWA. Denver, CO.

American Water Works Association. 2001. AWWA Standard for Polyacrylamide, ANSI/AWWA B453-01, Awwa. Denver, CO.

Amirtharajah, A. and S.C. Jones. 1996. Mixing for coagulation: Organic polymers, static mixers, and modeling. In Chemical Water and Wastewater Treatment IV, Edited by H. H. Hahn, E. Hoffmann and H. Odegaard. Pp.3-15, Berlin: Springer, Berlin.

Amy, G. L. and P.A. Chadik. 1983. Cationic polyelectrolytes as primary coagulants for removing trihalomethane precursors. J. Am. Water Works Assoc. 75, 527-531.

Anderson, N., B. Bolto, C. Chin. And L. Kolarik. 1993a. Recovery, purification and reuse of alum in coagulation processes. 15th Fed. Conv. Aust. Water & Wastewater Assoc. Vol. 4, p. 1053, Sydney, Australia.

Anderson, N. J., B.A. Bolto, R.J. Eldridge, and M.B. Jackson. 1993b. Polyampholytes for water treatment with magnetic particles. Reactive Polymers 19, 87-95.

Andrews, S.A.; V.Y. Taguchi. 2000. NDMA – Canadian Issues, Proceedings of Water Quality Technology Conference, American Water Works Association.

Arora, H., G. Digiovanniand M. Lechevallier. 2001. Spent filter backwash water contaminants and treatment strategies. J. Am. Water Works Assoc. 93, 100-112.

Attia, Y. A. and J. Rubio. 1975. Determination of very low concentrations of polyacrylamide and polyethyleneoxide flocculants by nephelometry. Br. Polym. J. 7, 135-138.

Baade, W., D. Hunkeler, and A.E. Hamielec. 1989. Copolymerization of acrylamide with cationic monomers in solution and inverse-microsuspension. J. Appl. Polym. Sci. 38, 185-201.

Bache, D. H., E. Rasool, D. Moffat, and M. Johnson 2000. Floc □abeling□ion□ion at full scale: The case of humic waters. In Chemical Water and Wastewater Treatment VI, Edited by H. H. Hahn, E. Hoffmann and H. Odegaard. P. 67, Springer, Berlin.

Barron, W., B.S. Murray, P.J. Scales, T.W. Healy, D.R. Dixon,. And M. Pascoe. 1994. The streaming current detector: a comparison with conventional electrokinetic techniques. Colloids and Surfaces A: Physicochemical and Engineering Aspects 88, 129-139.

Bartelt, A., D. Horn, W. Geiger, and G. Kern. 1994. Control and optimization of flocculation processes in the laboratory and in plant. Progr. Colloid & Polymer Sci. 95, 161-167.

Becker, N. S. C., N.A. Booker, A. Davey, S.R. Gray, R. Jago and C. Ritchie. 2000. The role of organic polyelectrolytes in high rate alternatives to primary separation. In Chemical Water and Wastewater Treatment VI, Edited by H. H. Hahn, E. Hoffmann and H. Odegaard. P. 223, Springer, Berlin.

Beckett, R. 1990. The surface chemistry of humic substances in aquatic systems. In Surface and Colloid Chemistry in Natural Waters and Water Treatment, Edited by R. Beckett. Pp. 3-20, Plenum, New York.

Beckett, R. and N.P. Le. 1990. The role of organic matter and ionic composition in determining the surface charge of suspended matter in natural waters. Colloids and Surfaces 44, 35-49.

Bennett, D. M., B.A. Bolto, D.R. Dixon, R.J. Eldridge, N.P.Le, and C.S. Rye. 2000. Determining the fate of flocculants by fluorescent tagging. Chemical Water and Wastewater Treatment VI, Edited by H. H. Hahn, E. Hoffmann and H. Odegaard. P. 159, Springer, Berlin.

Bernhardt, H. 1994. Rehabilitation for reduction of applied chemicals and removal of trace contaminants. In Proc. Regional Conf. p. 161, Int. Water Supply Assoc., Zurich.

Bernhardt, H. and H. Schell. 1996. Experience in coagulant control by use of a charge titration unit. J. Water SRT-Aqua 45, 19-27.

Biesinger, K. E. and G.N. Stokes. 1986. Effects of synthetic polyelectrolytes on selected aquatic organisms. J. Water Poll. Control Fed. 58, 207-213

Bilanovich, B., G. Shelef, and A. Sukenik. 1988. Flocculation of microalgae with cationic polymers – effects of medium salinity. Biomass 17, 65-76 1988.

Black, A. P., F.B. Birkner, and J.J.Morgan. 1966. The effect of polymer adsorption on the electrokinetic stability of dilute clay suspensions. J. Colloid and Interface Sci. 21, 626-648.

Bloys Van Treslong, J. C. and A. J. Staverman. 1974. Poly(ethyleneimine) II. Potentiometric titration behaviour in comparison with other weak polyelectrolytes Rec. Trav. Chim. Pays-Bas 93, 171-178.

Boggs, S., D.G. Livermore and M.G. Seitz. 1985. Humic macromolecules in natural waters. Rev. Macromol. Chem. Phys. C25, 599-657.

Bolto, B. A. 1995. Soluble polymers in water purification. Progress in Polymer Science, 20, 987-1041. Elsevier.

Bolto, B. A. and A.D. Levine, A. D. 2002. Reaction of polyelectrolytes with disinfectants to produce disinfection by-products. In Proc. IWA Water Congress, Melbourne. CD ROM.

Bolto, B. A., D.R. Dixon, R.J. Eldridge, and S.J. King.. 2001. Cationic polymer and clay for natural organic matter removal. Water Research 35, 2669-2676.

Bolto, B. A., D.R. Dixon, R.J. Eldridge, and S.J. King. 1998. The use of cationic polymers as primary coagulants in water treatment. In Chemical Water and Wastewater Treatment V, Edited by H. H. Hahn, E. Hoffmann and H. Odegaard. Pp.173-182, Springer, Berlin.

Burkert, H. 1970. The estimation of trace polyacrylamide in water. GWF-Wasser/Abwasser, Munich 111, 282-286.

Bursill, D. B., P.T. Hine, and J.Y. Morran. 1985. The effect on natural organics on water treatment processes. 11th Fed. Conv. Aust. Water & Wastewater Assoc., pp. 197-204, Sydney, Australia.

Cary, G. A., J.A. Mahon, and W.J. Kuc. 1987. The effect of suspended solids and naturally occurring dissolved organics in reducing the acute toxicities of cationic polyelectrolytes to aquatic organisms. Environ. Toxicol. And Chem. 6, 469-474.

Chamberlain, R. J. 1981. Polyelectrolyte makeup and handling. Polyelectrolytes for Water and Wastewater Treatment W. L. K. Schwoyer Ed., p. 243, CRC Press, Boca Raton.

Chang, E. E., P.C. Chiang, S.H. Chao, and C.H. Liang. 1999. Effects of polydimethyldiallylammonium chloride coagulant on formation of chlorinated by-products in drinking water. Chemosphere, 39, 1333-1346.

Chang, L. L., D.L. Raudenbush, and S.K. Dentel.. 2001. Aerobic and anaerobic biodegradability of a flocculant polymer. Water Sci. Tech. 44[2-3], 461-468.

Child, P., G. Kaar, D. Benitz, D., P. Fowlie, and R. Hong-You. 1991. Reaction between chlorine and a dimethylamine containing polyelectrolyte leading to the formation of N-nitrosodimethylamine. In Proc. Fourth National Conf. on Drinking Water, Ontario, pp. 58-71.

Coccagna, L. 1989. Direct filtration. In Water, Wastewater, and Sludge Filtration Edited by S. Vigneswaran and R. Ben Aim. P.57, CRC Press, Boca Raton.

Colbourne, J. S. 1989. Thames Water Utilities experience with Cryptosporidium. Proc. Am. Water Works Assoc. Water Quality Technology Conf., Denver, pp. 275-278.

Cornwall, D. A., M.J. MacPhee, N.E. McTigue, H. Arora, H., G. DiGiovanni, M. LeChevallier, and J.S. Taylor. 2001. Treatment options for Giardia, Cryptisporidium, and other contaminants in recycled backwash water. AWWA Research Foundation Report, Denver, Colo.:AWWARF and AWWA.

Criddle, J. 1990. A Review of the Mammalian and Aquatic Toxicity of Polyelectrolytes, Report NR 2545, Foundation for Water Research, Medmenham, UK.

Crozes, G., P. White, and M. Marshall, 1995. Enhanced Coagulation: Its Effect on NOM Removal and Chemical Costs Journal American Water Works Association 87,1, 78-89

Davis, M. K., S.E. Barrett, C.J. Whang, Y. Gou, Y. and S. Liang. 2000. N-nitrosodimethylamine in surface water. In Water Quality Technical Conf. Proc., AWWA, Denver.

Dentel, S.K., L.L. Cheng, D.L. Raudenbush, R.W. Junnier, M.M. Abu-Orf. 2000. Analysis and Fate of Polymers in Wastewater Treatment, Project 94-REM-2, Water Environment Research Foundation, Alexandria, VA

Dentel, S. K., M.M. Abu-Orf, and C.A.Walker. 2000. Optimisation of slurry flocculation and dewatering based on electrokinetic and rheological phenomena. Chem. Eng. J. 80, 65-72.

Dentel, S.K., M.M. Abu-Orf, and N.J. Griskowitz. 1993. *Guidance Manual For Polymer Selection in Wastewater Treatment Plants* Project 91-ISP-5, Water Environment Research Foundation.

Dentel, S. K., K.M. Wehnes, and M.M. Abu-Orf. 1994. Using streaming current and other parameters for polymer dose control in sludge conditioning. In Chem. Water Wastewater Treat. III, Proc. Gothenburg Symp. Edited by R. Klute and H. H. Hahn. Pp. 373, Springer, Berlin.

Dentel, S. K. and K.M. Kingery. 1989. Using Streaming Current Detectors in Water Treatment. Journal American Water Works Association, 81, 85-94.

Dentel, S.K. 1995. Use of the Streaming Current Detector in Coagulation Monitoring and Control. J.Water SRT-Aqua, 44, 70-79.

Dezinger, W., H. Hartmann, C. Gousetis, and K.H. Buechner. 1990. Use of water-soluble graft polymers of naturally occurring materials in water treatment. Ger. Offen. DE 4,038,908 assigned to BASF A.-G.

Durand, G., F. Lafuma, and R. Audebert. 1988. Adsorption of cationic polyelectrolytes at clay-colloid interface in dilute aqueous suspensions – effect of the ionic strength of the medium. Prog. Colloid Polym. Sci. 76, 278-282.

Durand-Piana, G., F. Lafuma, and R. Audebert. 1987. Flocculation and adsorption properties of cationic polyelectrolytes toward Na-montmorillonite dilute suspensions. J. Colloid Interface Sci. 119, 474-480.

Eades, A., B.J. Bates, and M.J. Macphee. 2001. Treatment of spent backwash water using dissolved air flotation. Water Sci. Technol. 43[8], 59-66.

Edzwald, J. K. 1993. Coagulation in drinking water treatment: particles, organics and coagulants. Water Sci. Technol. 27[11], 21-35.

Edzwald, J. K. 1986. Conventional water treatment and direct filtration: Treatment and removal of total organic carbon and trihalomethane precursors. Organic Carcinogens in Drinking Water N. M. Ram, E. J. Calabrese and R. F. Christman, Ed., p.208, Wiley, New York.

Edzwald, J. K., J.D. Haff, and J.W. Boak.1977. Polymer coagulation of humic acid waters. J. Env. Eng. Divn. 103, 989-1000.

Eriksson, L. and B. Alm. 1993. Characterization of activated sludge and conditioning with cationic polyelectrolytes. Water Sci. Tech. 28[1], 203-212.

Eriksson, L., B. Alm and P. Stenius. 1993. Formation and structure of polystyrene latex aggregates obtained by flocculation with cationic polyelectrolytes 1. Adsorption and optimum flocculant concentrations. Colloids and Surfaces A. Physicochemical and Engineering Aspects, 70, 47-60.

Farrow, J. B., P.D. Fawell, R.R. Johnston, T.B. Nguyen, M. Rudman, K. Simic, and J.D.Swift. 2000. Recent developments in techniques and methodologies for improving thickener performance. Chem. Eng. J. 80, 149-155.

Faust, A. D. and O.M. Aly. 1983. Chemistry of Water Treatment, Butterworths, Boston, pp.326-328.

Feige, M. A., E. M. Glick J. W. Munch, D. J. Munch, R. L. Noschang and H. J. Brass 1980. Potential contaminants introduced into water supplies by the use of coagulant aids. In Water Chlorination: Environmental Impact and Health Effects, Vol. 3, Edited by R. L Jolley, W. A. Brungs, R. B. Cumming and V. A. Jacobs, Ann Arbor Sci. Pub., Ann Arbor, Michigan, pp. 789-799.

Fielding, M., J. Hutchison, D. M. Hughes, W. H. Glaze and H. S. Weinberg 1999. Analytical methods for polymers and their oxidative by-products. AwwaRF Project Report, Denver, Colo: AwwaRF and AWWA.

Fiessinger, F., J. Mallevialle, and A. Bruchet. 1983. Fate of polymers in the treatment process, Proc. Seminar on Use of Organic Polymers in Water Treatment, 37-50, Am. Water Works Assoc. Res. Div., Denver.

Ghosh, M. M., C.D. Cox, and T.M. Prakash. 1985. Polyelectrolyte selection for water treatment. J. Am. Water Works Assoc. 77[3], 67-73.

Gibbs, R. J.1983. Effect of natural organic coatings on the coagulation of particles. Environ. Sci. Technol., 17, 237-240.

Goodrich, M. S., L.H. Dulak, M.A. Freedman, and J.J. Lech. 1991. Acute and long term toxicity of water-soluble cationic polymers to rainbow trout Oncorhynchus mykiss. And the modification of toxicity by humic acid. Environ. Toxicol. Chem. 10[4], 509-515.

Goppers, V. and C.P. Straub. 1976. Polyelectrolyte persistence in a municipal water supply. J. Am. Water Works Assoc. 68, 319-321.

Graham, J. E., S.A. Andrews, G.J. Farquhar, and O. Meresz. 1996. Thiram as an NDMA precursor in drinking water treatment, Proc. Am. Water Works Assoc. Water Quality Technology Conf., Toronto, Vol. C, pp. 15-27.

Gregory, J. 1985a. The action of polymeric flocculants. In Flocculation, Sedimentation and Consolidation Edited by B. M. Moudigil and P. Somasundaran. P.125, Engineering Foundation, Washington.

Gregory, J. 1985b. Turbidity fluctuations in flowing suspensions. J. Colloid & Interface Sci. 105, 357-372.

Gregory, J. and D.W. Nelson. 1986. Monitoring of aggregates in flowing suspensions. Colloids and Surfaces 18, 175-188.

Gregory, J. and S.Y. Lee. 1990. The effect of charge density and molecular mass of cationic polymers on flocculation kinetics in aqueous solution. J. Water SRT-Aqua 39, 265-274.

Gutcho, S. 1977. Waste Treatment with Polyelectrolytes and Other Flocculants, Noyes Data, Park Ridge, New Jersey.

Haarhoff, J. and J.L. Cleasby. 1989. Direct filtration of Chlorella with cationic polymer. J. Environ. Eng. 115, 348-366.

Hall, W. S. and R.J. Mirenda. 1991. Acute toxicity of wastewater treatment polymers to Daphnia pulex and the flathead minnow Pimephales promelas. And the effects of humic acid on polymer toxicity. Research J. Water Poll. Control Fed. 63, 895-899.

Hall, T. and R.A. Hyde. 1992. Water Treatment Processes and Practices, Water Research Centre, Swindon, UK.

Hamilton, J. D., K.H. Reinhert, and M.B. Freeman. 1994. Aquatic risk assessment of polymers. Environ. Sci. Technol. 28, 187A-192A.

Hanasaki, T., H. Ohnishi, A. Nikaidoh, S. Tanada, and K. Kawaski. 1985. Determination of Trace Polymer in Wastewater" Bull. Environ. Contam. Toxicol, 35, 476-481.

Harrington, A. and P. Engelhardt. 1979. Cationic starches as industrial flocculants.

International Water Conference, p. 313, Eng. Soc. W. Penn., Pittsburg.

Hogg, R. 2000. Flocculation and dewatering. Internat. J. Min. Proc. 58, 1-4.

Hollander, A. F., P. Somasundaran, and C.C. Gryte. 1981. Adsorption characteristics of polyacrylamide and sulfonate-containing polyacrylamide copolymers on sodium kaolinite. J. Appl. Polym. Sci. 26, 2123-2138.

Hongeveen, N. G., M. A. Cohen Stuart and G. J. Fleer. 1996. Can charged block co.polymers act as stabilizers and flocculants of oxides? Colloids & Surfaces A. Physicochemical and Engineering Aspects 117, 77-88.

Hoog, R. 1984. Collision Efficiency Factors for Polymer Flocculation, Journal of Colloid and Interface Science, 102, 1, 232-236

Hoover, M. F., 1970. Cationic quaternary polyelectrolytes – a literature review. J. Macromol. Sci.- Chem. A4, 1327-1417.

Howes, N. J. and G. Harper. 1998. Review of analytical methodology for cationic polyacrylamides at levels of environmental significance. R&D Technical Report No. E47, Environment Agency, Bristol, UK, 34 pp.

Hozalski, R.M. and H. Li. 1999. Development of an Innovative Method for Natural Organic Matter Characterization Using Capillary Electrophoresis. Proc. AWWA Water Quality Technology Conf.

Huang, C. P., S.C. Chen, and J.R. Pan. 2000. Optimal condition for modification of chitosan: A biopolymer for coagulation of colloidal particles. Water Research 34, 1057-1062.

Huang, H.-J. and H.H. Yeh. 1993. Organic fractionation for water treatment process evaluation. Proc. Water Tech. Conf., Pt I, p.257, Am. Water Works Assocn., Denver.

Hutchison, P. R. and T.W. Healy. 1990. Coagulation and flocculation – destabilizing practices? With particular reference to metal ion coagulants. Surface and Colloid Chemistry in Natural Waters and Water Treatment R. Beckett, Ed., p.119, Plenum, New York.

Jackson, G. E. 1980. Granular media filtration in water and wastewater treatment: Part II. CRC Critical Reviews in Environmental Control 11, 1-36.

Jiang, J.-Q., N.J.D. Graham, and C. Harward. 1993. Comparison of polyferric sulphate with other coagulants for the removal of algae and algal-derived organic matter. Water Sci. Tech. 27[11], 221-230.

Jobb, D. B., R.B. Hunsinger, O. Meresz and V.Y. Taguchi. 1992. A study of the occurrence and inhibition of formation of N-nitrosodimethylamine in the Ohsweken water supply. Proc. Fifth National Conf. on Drinking Water, Winnipeg, pp. 241-252.

Johannsen, K., M. Assenmacher, M. Kleiser, G. Abbt-Braun, H. Sontheimer, and F.H. Frimmel. 1993. Influence of the molecular size on the adsorption of humic compounds. Vom Wasser 81, 185-196.

Kan, C. C. and G.S. Huang. 1998. Coagulation monitoring in surface water treatment facilities. Water Sci. Tech. 38[3], 237-244.

Kaiser, K. L. E. and J. Lawrence. 1977. Polyelectrolytes: potential chloroform precursors. Science 196, 1205-1206.

Kawabata, N., D. Uchihori, S. Fukuda, and H. Funahashi. 1994. Digestion of crosslinked poly vinylpyridinium halide. By activated sludge, and application to make poly methyl methacrylate. Biodegradable by incorporation of a pyridinium group into the main chain. Appl. Polym. Sci. 51, 33-42.

Kawamura, S. 1976. Considerations on improving flocculation. J. Am. Water Works Assoc. 68, 328-336.

Keenan, H. E., E.N. Papavasilopoulos, and D.H. Bache. 1998. Measurement of polymer residuals in an alum sludge. Water Research 32, 3173-3176.

Kenefick, S. L., S.E. Hrudey, H.G. Peterson, and E.E. Prepas. 1993. Toxin release from Microcystis aeruginosa after chemical treatment. Water Sci. Tech. 27[3-4], 433-440.

Kim, C. H., J.W. Choi, H.J. Chun, and K.S. Choi. 1997. Synthesis of chitosan derivatives with quaternary ammonium salt and their antibacterial activity. Polymer Bull. 38, 387-393.

Kirie, K., T. Zaitsu, and C. Igarashi. 1986. Determination of cationic organic polymeric flocculant. Jpn. Kokai Tokkyo Kohu JP 86 86,656 assigned to Ebara-Infilco Company Ltd.

Kohut, K. D., and Andrews, S. A. 2003. Polyelectrolytes as amine precursors for *N*-nitrosodimethylamine in drinking water. Proc.Am. Water Works Assoc. Ann. Conf., Anaheim, California.

Korshin, G.V., C-W. Li, and M.M. Benjamin 1997. Monitoring the properties of natural organic matter through UV spectroscopy: A consistent theory. Water Research, 31, 7, 1787-1795

Kruger, D. and W. Gedney. 2000. Case studies of N-nitrosodimethylamine in groundwater supplies. Water Quality Technical Conf. Proc., AWWA, Denver.

Kuo, C-Y. and T. Provder. 1987. An Overview of Size Exclusion Chromatography for Polymers and Coatings. American Chemical Society Symposium Series 352: Detection and Data Analysis in Size Exclusion Chromatography, 2-28 , American Chemical Society, Washington D.C.

Lafuma, F. and G. Durand. 1989. 13C NMR spectroscopy of cationic coplymers of acrylamide. Polymer Bulletin 21, 315-318.

Lancaster, J. E., L. Bacchei, L. and H.P. Panzer.. 1976. The structure of poly diallyldimethylammonium chloride. By 13C-NMR spectroscopy. J. Polym. Sci. B14, 549-554.

Lartiges, B. S., J.Y. Bottero, C. Democrate, and J.F. Coupel.1995. Optimising polymer demand by following floc size distribution. J. Water SRT-Aqua 44, 219-223.

Lawrence, J. 1989. Analysis of Trace Organics in the Aquatic Environment B. K. Afghan and A. S. Y. Chau, Ed., p.313, CRC Press, Boca Raton.

LeChevallier, M., W.D. Norton, and R.G. Lee. 1991. Giardia and Cryptosporidium spp. in filtered drinking water supplies. Applied & Environ. Microbiol. 57, 2617-2621.

Lee, S. H., W.S. Shin, M.C. Shin, S.J. Choi, and L.S. Park. 2001. Improvement of water treatment performance by using polyamine flocculants. Environ. Technol. 22, 653-659.

Lee, S. Y. and J. Gregory. 1991. The effect of charge density and molecular mass of cationic polymers on flocculation kinetics in aqueous solution. Water Supply 9, 11-17.

Lenz, R. W. 1993. Biodegradable polymers. Adv. Polym. Sci. 107, 1-40.

Letterman, R. D. and R.W. Pero. 1990. Contaminants in polyelectrolytes used in water treatment. J. Am. Water Works Assoc. 82[11], 87-97.

Letterman, R.D. and P.Sricharoenchaikit 1982 Interaction of Hydrolyzed Al and Polyelectrolyte Coagulants *Journal Environmental Engineering Division*, ASCE, 108, EE5, 883-899.

Leu, R.-J. and M.M. Ghosh. 1988. Polyelectrolyte characteristics and flocculation. J. Am. Water Works Assoc. 80[4], 159-167.

Levine, A. D., R. L. Swickley, J. R. Carman, C. M. Ballif, M. S. Beal and M. E. Owens 1998. Permanganate interactions with cationic polymer and NOM in surface water treatment, In Proc. Am. Water Works Assoc. Water Quality Technology Conf., Denver.

Levine, A. D., L.M. Mercurio, and J.R. Carman. 1996. Coagulation of natural organic matter from surface water using cationic polymer. In Chemical Water and Wastewater Treatment IV Edited by H. H. Hahn, E. Hoffmann and H. Odegaard. P.189, Springer, Berlin.

Levine, N. M. 1981. Natural polymer sources. Polyelectrolytes for Water and Wastewater Treatment W. L. K. Schwoyer Ed., p. 47, CRC Press, Boca Raton.

Lockyear, C. F., P.J. Jackson, and J.H. Warden. 1983. Polyelectrolyte Users' Manual, Technical report TR 184, Water Research Centre, Swindon, UK.

Logsdon, G. S., D.G. Neden, M.D. Ferguson, and S.D. LaBonde. 1993a. Testing direct filtration for the treatment of high-turbidity water. J. Am. Water Works Assoc. 85[12], 39-46.

Logsdon, G. S., D.G. Neden, and S.D. LaBonde. 1993b. Direct filtration of turbid water. Proc. Water Quality Conf., p. 679, Am. Water Works Assoc., Denver.

Low, K. C. and L.P. Koskan. 1993. Synthetic polyaspartic acid and its uses. Abstracts of Papers, 206th National Meeting, PMSE 124, Am. Chem. Soc., Washington.

Lurie, M. and M. Rebhun. 1997. Effect of properties of polyelectrolytes on their interaction with particulates and soluble species. Water Sci. Tech. 36[4] 93-101.

Luttinger, L. B. 1981. The use of polyelectrolytes in filtration processes. Polyelectrolytes for Water and Wastewater Treatment W. L. K. Schwoyer Ed., p. 211, CRC Press, Boca Raton.

MacPhee, M.J., and D.A. Cornwell. 2002. Trace contaminants in drinking water chemicals, AWWARF Project Report, Denver, Colo: AWWARF and AWWA.

McKague, A. B. 1974. Flocculating agents derived from kraft lignin. J. Appl. Chem. Biotechnol. 24, 607-615.

Mallevialle,J., A. Bruchet and F. Fiessinger 1984. How safe are organic polymers in water treatment? J. Am. Water Works Assoc. 76[6], 87-93.

Matsumura, S., J. Takahashi, S. Maeda, and S. Yoshikawa. 1998. Molecular design of biodegradable polyelectrolytes. II. Biodegradation of poly sodium vinyloxyacetate. And its copolymers. Kobunshi Ronbunshu 45, 325-331.

Meister, J. J. and C.T. Li. 1990. Cationic graft cop[olymers of lignin as sewage sludge dewatering agents. Polym. Prepr. Am.Chem. Soc., Div. Polym. Chem. 31[1], 664.

Metcalf and Eddy. 2003. Wastewater Engineering: Treatment and Reuse Fourth Edition, edited by G. Tchobanoglous, F.L. Burton, and H.D. Stensel, McGraw-Hill.

Mitch W.A. and D.L. Sedlak. 2002. Factors affecting the formation of NDMA during chlorination. Environ. Sci. Technol., 36, 588-595

Mohammed, A., S. Weir, S. and G. Moody. 2000. Flocculants: Development trends. Filtration and Separation 37[8], 24-27.

Muzzarelli, R. A. A. 1977. Chitin, p. 95, Pergamon Press, Oxford.

Narita, T., R. Ohtakeyama, M. Matsukata, J.P. Gong, and Y. Asada,. 2001. Kinetic study of cell disruption by ionic polymers with varied charge density. Colloid Polym. Sci. 279, 178-183.

Narkis, N. and M. Rebhun. 1983. Inhibition of flocculation processes in systems containing organic matter. J. Water Poll. Control Fed. 55, 947-955.

Narkis, N., B. Ghattas, M. Rebhun, and A.J. Rubin. 1991. The mechanism of flocculation with aluminium salts in combination with polymeric flocculants as flocculant aids. Water Supply 9, 37-44.

NSF International 2001. Certified product listings ANSI/NSF Standard 60. www.nsf.org.

NSF/ANSI STANDARD 60: Drinking Water Treatment Chemicals – Health Effects http://www.nsf.org/Certified/PwsChemicals/

Ndabigengesere, A., K.S. Narasiah, and B.G. Talbot. 1995. Active agents and mechanism of coagulation of turbid water using Moringa oleifera. Water Research 29, 703-710.

Nilsson, R. and M. Kvant. 1986. Developments of new chemicals in order to reduce dosage and prevent high salt residuals. In Recycling in Chemical Water and Wastewater Treatment Edited by H. H. Hahn, R. Klute and P. Balmer. P.323, Schriftenreihe des ISWW, Universität Karlsruhe.

Njam, I. and R.R. Trussell. 2001. NDMA formation in water and wastewater, J. Am. Water Works Assoc. 93[2], 92-99.

Norden, M., and E. Dabek-Ziotorzynska. 1997. Characterization of humic substances using capillary electrophoresis with photodiode array and laser-induced fluorescence detection. Electrophoresis, 18, 292-299.

Norman, P. I. and R. Seddon. 1991. Pollution control in the textile industry – the chemical auxiliary manufacturer's role. J. Soc. Dyers Colour., 107, 215-219.

Novak, J. T. and N. Bandak. 1994. The effect of shear on the dewatering of water treatment residuals. J. Am. Water Works Assoc. 86[11], 84-91

Nozaic, D. J., S.D. Freese, and P. Thompson. 2001. Long term experience in the use of polymeric coagulants at Umgeni Water. Water Sci. Technol.: Water Supply 1[1], 43-50.

National Sanitation Foundation 2004. Certified product listings ANSI/NSF Standard 60. www.nsf.org.

Parazak, D. P., C.W. Burkhardt, and K.J. McCarthy. 1987. Determination of low levels of cationic polyelectrolytes in water. Anal. Chem. 59, 1444-1445.

Parazak, D. P., C.W. Burkhardt, K.J. McCarthy, and M.P. Stehlin. 1988. Hydrophobic flocculation. J. Colloid and Interface Sci., 123, 59-72.

Pariser, E. R. and D.P. Lombardi. 1989. Chitin Sourcebook: A Guide to the Research Literature, p. II: L – 1, Wiley, New York.

Petruzzelli, D., A. Volpe, A.C. Di Pinto, and R. Passino. 2000. Conservative technologies for environmental protection based on the use of reactive polymers, Reactive Polymers, 45, 95-10.

Pompe, S., K-H. Heise, and H. Nitsche. 1996. Capillary electrophoresis for a finger-print characterization of fulvic and humic acids" Journal of Chromatography A, 723, 1996, 215-218.

Rebhun, M., Z. Fuhrer, and A. Adin. 1984. Contact flocculation-filtration of humic substances. Water Research 18, 963-970.

Rebhun, M. and M. Lurie.1993. Control of organic matter by coagulation and floc separation. Water Science & Technology, 27[11], 1-20.

Reuter, J. M. and A. Landscheidt. 1988. Polyelectrolytes for the treatment of tap and filter backwashing water. In Pretreatment in Chemical Water and Wastewater Treatment Edited by H. H. Hahn and R. Klute. P. 96, Springer-Verlag, Berlin.

Rice, D. M., J.H. Denysschen, and G.J. Stander. 1964. Evaluation of a Tannin-base Polyelectrolyte as a Coagulant for Turbid Waters, CSIR Research Report 223, Pretoria, South Africa.

Rigol, A., J.F. Lopez-Sanchex, and G. Rauret. 1994. Capillary zone electrophoresis of humic acids. Journal of Chromatography A., 664, 301-305.

Rose, J. B., A. Cifrano, M.S. Madore, C.P.cGerba, C.R. Sterling, and M.J. Arrowood. 1986. Detection of Cryptosporidium from wastewater and freshwater environments. Water Sci. Technol. 18[10], 233-239.

Rout, D., R. Verma, and S.K. Agarwal. 1999. Polyelectrolyte treatment – an approach for water quality improvement. Water Sci. Technol. 40[2], 137-141.

Satyanarayana, D. and P.R. Chatterji. 1993. Biodegradable polymers: Challenges and strategies. Rev. Macromol. Chem. Phys. C33, 349-368.

Schlauch, R. M. 1981. Coagulation for gravity type clarification andd thickening. Polyelectrolytes for Water and Wastewater Treatment W. L. K. Schwoyer Ed., p.91, CRC Press, Boca Raton.

Schwoyer, L. K. 1981. Sludge dewatering. Polyelectrolytes for Water and Wastewater Treatment W. L. K. Schwoyer Ed., pp. 159, CRC Press, Boca Raton.

Scott J. P., P.D. Fawell, D.E. Ralph, and J.B.Farrow. 1996. The shear degradation of high-molecular-weight flocculant solutions. J. Applied Polymer Sci. 62, 2097-2106.

Senthilnathan, P. R. and R.G. Sigler. 1993. Improved sludge dewatering by dual polymer conditioning. Water Sci. Technol. 28[1], 53-57.

Singer, P.C. and G.W. Harrington. 1993. Coagulation of DBP precursors: theoretical and practical considerations. Proc. Water Technology Conf., Pt. I, p.1, Am. Water Works Assoc., Denver 1993.

Smith-Palmer, T., N. Campbell, J.L. Bowman, and P. Dewar. 1994. Flocculation behavior of some cationic polyelectrolytes. J. Appl. Poyml. Sci. 52, 1317-1325.

Soponkanaporn, T. and R. Gehr 1989. The degradation of polyelectrolytes in the environment: Insights provided by size exclusion chromatography measurements. Water Sci. Technol. 21, 857-868.

Stockham, P. and J. Morran. 2000. Potential disinfection by-products from poly DADMAC. Based polyelectrolytes. Proc. WaterTECH Conference, Aust. Water Assocn., Sydney.

Stumm, W. 1992. Chemistry of the Solid-Water Interface, p.114, Wiley, New York.

Tanaka, H. and L. Odberg. 1989. Preparation of cationic polyacrylamides by a modified Hofmann reaction: Fluorescent □abeling of cationic polyacrylamides. J. Polym. Sci. Part A: Polym. Chem. 27, 4329-4339.

Tanaka, H. and Y. Sakamoto. 1993. Polyelectrolyte titration using fluorescent indicator. I. Direct titration of anionic and cationic polyelectrolytes with 10-4N standard solutions. J. Polym. Sci. Part A: Polym. Chem. 31, 2687-2691.

Tarquin, A. J., G.B. Haan, and D. Rittmann. 1998. Polymer cost and performance evaluation. Water Engineering & Management 145[1] 27-31.

Taylor, K. C. and H.A. Nasr-El-Din. 1994. Acrylamide copolymers: A review of methods for the determination of concentration and degree of hydrolysis. Petroleum Sci. & Engineering 12, 9-23.

Terrell, D. L. 1977. Organic polymers replace alum and improve water quality in Ithaca. J. Am. Water Works Assoc. 69, 263-266.

Thurman, E. M. 1985. Organic Geochemistry of Natural Waters, p.3, Martinus Nijhoff-Dr Junk Publishers, Dortrecht.

Timofeeva, S. S., A.M. Beim, and A.A. Beim. 1994. Ecologo-technological principles of the choice of flocculants for wastewater purification from clay suspensions. Khimiya i Teknologiya Vody 16, 72-76.

Tipping, E. 1993. Model binding by humic acids. Colloids and Surfaces A: Physicochemical and Engineering Aspects 73, 117-131.

Tiravanti, G., F. Lore, and G. Sonnante. 1985. Influence of the charge density of cationic polyelectrolytes on sludge conditioning. Water Res. 19, 93-97.

Tryby, M. E., R.J. Miltner, and R.S. Summers. 1993. TOC removal as a predictor of DBP control with enhanced coagulation. Proc. Water Quality Technology Conf., Pt. I, p. 21, Am. Water Works Assoc., Miami. Denver.

U.S. EPA , 2002. Technical Factsheet on: Epichlorhydrin. http://www.epa.gov/OGWDW/dwh/t-voc/epichlor.html

U.S. EPA , 2004. Technical Factsheet on: Acrylamide. http://www.epa.gov/OGWDW/dwh/c-voc/acrylami.html

Valentine, R.L., J. Choi, Z. Chen, S.E. Barrett, C. Hwang, C., Y.C. Guo, M. Whener, and S. Fitzsimmons, 2004. Factors affecting the formation of NDMA in water and occurrence, AwwaRF

Van der Steeg H. G. M., M.A. Cohen-Stuart,.A., de Keizer, and B.H. Bijsterbosch, 1992. Polyelectrolyte adsorption: A subtle balance of forces. Langmuir 8, 2538-2546.

Van de Wetering, P., N.J. Zuidam, M.J. van Steenbergen, O.A. van der Houwen, G. J., Underberg, W. J. M. and W. E. Hennink, 1998. A mechanistic study of poly 2-dimethylamino.ethyl methacrylate. Macromolecules 31, 8063-8068.

Van Leeuwen, J., H. Schell, M. Berger, M. Drikas, D. Bursill, C. Chow, and J. Clasen. 1997. Comparison of coagulant doses determined using a charge titration unit with a jar test procedure for eight German Surface waters, J Water SRT, Aqua, 46, 5, 261-273.

Veal, C. R. 1990. Application technology of the streaming current detector. Ultrapure Water 7[4], 20-23.

Vigneswaran, S. and R. Ben Aim. 1989. Overview of deep bed filtration:different types and mathematical models. Water, Wastewater, and Sludge Filtration S. Vigneswaran and R. Ben Aim, Ed., p. 1, CRC Press, Boca Raton.

Vik,E. A. and B. Eikebrokk. 1989. Coagulation process for removal of humic substances from drinking water. Aquatic Humic Substances I. H. Suffet and P. MacCarthy Ed., p.385, Am. Chem. Soc., Washington.

Vitvitskaya, B. R., A.A. Korolev, J.N. Skachkova, G.A. Savonicheva, S.G. Sergeev, and O.L. Nokova. 1988. Validation of maximum allowable concentration of polydimethyldiallylammonium chloride in water reservoirs. Gig. Sanit. 3, 66-68 1988.

Vorchheimer, N. 1981. Synthetic polyelectrolytes. Polyelectrolytes for Water and Wastewater Treatment W. L. K. Schwoyer Ed., p. 1, CRC Press, Boca Raton.

Vostrcil, J. and F. Juracka.1976. Commercial Organic Flocculants, Noyes Data, Park Ridge, New Jersey.

Voycheck, C. L. and J.S. Tan.. 1993. Ion-containing polymers and their biological interactions. Polyelectrolytes: Science and Technology M. Hara Ed., p. 309, Dekker, New York.

Walzer, J. G. 1981. Polyelectrolytes and coagulants for the flotation process. Polyelectrolytes for Water and Wastewater Treatment W. L. K. Schwoyer Ed., p. 145, CRC Press, Boca Raton.

Wan Ngah, W. S. and A. Musa. 1998. Adsorption of humic acid onto chitin and chitosan. J. Appl. Polymer Sci. 69, 2305-2310.

Wandrey, C. and W. Jaeger. 1984. Synthesis of polyelectrolytes. 6. Copolymerization of dimethyldiallylammonium chloride and acrylamide. Acta Polym. 36, 100-102.

Wang, L. K. and W.W. Shuster. 1975. Polyelectrolyte determination at low concentrations. Ind. Eng. Chem., Prod.Res. Dev. 14, 312-314.

Wang, T. K. and R. Audebert. 1987. Flocculation mechanisms of a silica suspension by some weakly cationic polyelectrolytes. J. Colloid and Interface Sci. 119, 459-465.

Wang, J. 1999. A New Way to Optimize Polymer Dosage. Operations Forum, 12, 22-24.

Wilczak, A. J., A. Assadi-Rad, H. H. Lai, L. L. Hoover, J. F. Smith, R. Berger, F. Rodigari, J. W. Beland, L. J. Lazzelle, E. G. Kincannon, H. Bker and C. T. Heaney. 2003. Formation of NDMA in chloraminated water coagulated with DADMAC polymer. J. Am. Water Works Assoc. 95[9], 94-106.

WRc 1998. A review of polyelectrolytes to identify priorities for EQS development. R&D Technical Report P21, Foundation for Water Research, Marlow, UK.

Yokoi, H., T. Aratake, S. Nishio, J. Hirose, S. Hayashi, and Y. Takasaki. 1998. Chitosan production from Shoku distillery wastewaters by funguses. J. Fermentation and Bioengineering 85, 246-249.

Yeh, H.H. and M.M. Ghosh. 1981. Selecting Polymers for Direct Filtration Journal of American Water Works Association 73,4, 211-218.

Zhu, H., D.W. Smith, H.Zhou, and S.J. Stanley. 1996, Improving Removal of Turbidity Causing Materials by Using Polymers as a Filter Aid, Water Research 30, 1, 103-114.

ABBREVIATIONS

μg/L	microgram per liter
μS/cm	microsiemens per centimeter
Al	Aluminum
amu	atomic mass units
APAMs	Anionic polyacrylamides
ASP	active server pages
AWWA	American Water Works Association
AwwaRF	Awwa Research Foundation
$CaCO_3$	Calcium carbonate
CAS	Chemical Abstracts Service
CD	Charge density (CD)
CE	Capillary Electrophoresis
CHES	2-[N-cyclohexylamino]-ethane sulfonate
COD	Chemical oxygen demand
CPAMs	Cationic polyacrylamides
CPSTP	Continuous Pilot Scale Treatment Plant
DBPs	Disinfection byproducts
DE	Diatomaceous earth
DOC	Dissolved organic carbon
EPI-DMAs	Epichlorohydrin dimethylamine polyamine
Fe	Iron
FSCE	Free solution capillary electrophoresis
g	Gram
GAC	Granular Activated Carbon
HAAs	Haloacetic Acids
HANs	Haloaceto nitriles
HPLC/MS	High pressure liquid chromatography with mass spectrometry
ID	Internal diameter
KCl	Potassium chloride
kPa	Kilopascals
kV	Kilovolts
MCL	Maximum contaminant level
MCLG	Maximum contaminant level goal
MECC	Micellar electrokinetic capillary chromatography

meq	milliequivalent
MGD	Million gallons per day
mg/L	milligram per liter
min	minute
MS2	Male specific coliphage (virus that infects coliform bacteria)
mV	millivolts
MW	Molecular weight
NaCl	Sodium chloride
NaOH	Sodium hydroxide
NDMA	N-nitrosodimethylamine
nm	nanometers
NMR	Nuclear magnetic resonance
NOM	Natural organic matter
NTU	Nephelometric turbidity units
OD	Outer diameter
PAC	Powdered Activated Carbon
PAMs	Polyacrylamides
PCU	Platinum Color Units
PDMAEMA	Polydimethylaminoethyl methacrylate
PEI	Polyethyleneimine (PEI)
pK_a	Negative log of the Acid Dissociation Constant
polyDADMACs	Poly(diallyldimethyl-ammonium chloride)
RCOOH	Carboxyllic Acid functional group
SCD	Streaming Current Detector
SDWA	Safe Drinking Water Act
SEC	Size Exclusion Chromatography
TFF	Tangential flow filtration
THM	Trihalomethanes
TOC	Total Organic Carbon
TT	Treatment Technique
TTHM	Total trihalomethanes
Turb	turbidity
UV	Ultraviolet
VIS	visible light

IWA Publishing's authorised EU representative for General Product Safety Regulations is Diane D'Arras, 15 rue Duret, 75116 Paris, France, e-mail: safety@iwap.co.uk.

Printed and bound by CPI Group (UK) Ltd, Croydon, CR0 4YY
17/09/2026
02218802-0017